Green

Green Living

Practical ways to make your home environment friendly

Bernadette Vallely, Felicity Aldridge and Lorna Davies

Thorsons
An Imprint of HarperCollins*Publishers*

Thorsons
An Imprint of GraftonBooks
A Division of HarperCollins*Publishers*
77–85 Fulham Palace Road,
Hammersmith, London W6 8JB

Published by Thorsons
1 3 5 7 9 10 8 6 4 2

© 1991 Bernadette Vallely, Felicity Aldridge
and Lorna Davies

Bernadette Vallely, Felicity Aldridge and Lorna Davies
assert the moral right to be identified as
the authors of this work

A CIP catalogue record for this book
is available from the British Library

ISBN 0-7225-2470-6

Typeset by G&M, Raunds, Northamptonshire
Printed in Great Britain by Mackays of Chatham, Kent

All rights reserved. No part of this publication may be
reproduced, stored in a retrieval system, or transmitted,
in any form or by any means, electronic, mechanical,
photocopying, recording or otherwise, without the prior
permission of the publishers.

Contents

Acknowledgements 8
Introduction 9

1 Building and design 17
2 Power and energy 29
3 Decorating and DIY 45
4 Flooring and fabrics 59
5 Cleaning 73
6 Food and drink 89
7 Equipment 103
8 Junk and waste disposal 119
9 Hobbies and home crafts 133
10 Furniture 148
11 Medicine 160
12 Gardening 176

Further reading 193
Address list 195
Index 203

About the Authors

Bernadette Vallely is the director of the Women's Environmental Network. She has written and co-authored many books about the environment including the best-selling *1001 Ways to Save the Planet* (Penguin 1990), *The Young Person's Guide to Saving the Planet* (Virago 1990) and *The Sanitary Protection Scandal* (WEN 1989). Bernadette has been campaigning on social and environmental issues for nearly ten years. She lives with her partner, also an environmentalist, in North London.

Felicity Aldridge works at the London Ecology Centre where she is currently coordinating a new building project for the centre and its tenants. She also studies art psychotherapy at Goldsmiths College and has written a book about feminine spirituality. Felicity is very interested in linking the way we live our lives with the harmony of the earth. She is on the Board of Directors of the Women's Environmental Network.

Lorna Davies lives in Basildon, Essex where she works as a school nurse. She became concerned about the state of the environment after the birth of her two children (now aged four and five) and has been an active environmental campaigner for several years, She is on the Board of Directors of the Women's Environmental Network and this is her first book.

The Women's Environmental Network

All the royalties from this book go to the Women's Environmental Network (WEN), a non-profit organization campaigning and educating women who care about the environment. WEN is funded by memberships, the sale of books and other goods, as well as through grants from charitable bodies.

The Women's Environmental Network runs an information service to give clear, practical advice to the public who would like to know about lifestyle issues including sanitary protection, clothing, packaging, babies and children, chemicals and green consumerism.

WEN members are invited to give their support by joining local activist groups or supporting fundraising initiatives. Members receive a quarterly newsletter, keeping them informed of current campaigns and news. Women and men are equally welcome to join, but the organization is woman-centred.

This book was inspired by an exhibition called *The Green Home* which WEN coordinated during 1990. A full information pack about WEN's work is available from:

The Women's Environmental Network
287 City Road
London, EC1V 1LA

Acknowledgements

We would like to express our thanks and love for all the help and support that we have received, especially from the following people: Karen Milliard for stepping in at the last moment with so much energy and constructive help. Charo Lamas, Funmi Shobo, Luz and Frances Balfour for all the typing and computer work. To Ann Link for all her technical and scientific support. To all the women at WEN for their help and support. To Stewart for his love and energy (sic!). To Tom and Joyce for helping and putting up with Lorna and to Alex and Helen for being so good. To Paul for his patience and lifts home.

We would also like to express our grateful thanks to those organizations, authors, researchers and individuals who helped us to prepare the manuscript and answered our questions about a variety of environmental issues.

Introduction

We have now officially entered the Earth Day Decade, and a media-stirred multitude of citizens is asking what can be done, indeed, what they can do, about the environmental crisis.

Stephanie Mills

This book aims to explain some of the questions, and solutions, that homeowners have recently been asking about the environment. What cleaners should they use? Which paints are environmentally friendly? How can you save money and the environment in your home? What sort of lifestyle is possible and enjoyable? Solving some of these problems might be easier than we think, and access to information is generally a major factor in helping people make changes in their lifestyle.

What we do in our everyday life is unequivocally connected to the destruction of our planet and our environment. The mass production and use of chemicals and energy in manufacturing processes is due both to consumer demand, and the lack of attention paid by governments and individuals to the practices of the manufacturing companies.

Living a greener lifestyle means much more than buying a 'green product': it means a re-evaluation of our whole lives. This includes how we interact with our surroundings and the local community. How we act in our homes also affects the rest of the planet; for example, the act of buying a product made in the Third World brings into consideration the conditions of workers in that country, the pollution caused, the fuel it took to bring the product to the market country, and how we will dispose of it when we have finished with it. This is the 'cradle to grave' approach, which we must now adopt in order to get a true picture of the destruction we are causing.

For us individually, green living means having a healthy body, mind and spirit. Without these we cannot possibly live our lives to the full. The present situation means that the human race may die out and become extinct because of pollution and destruction. Is that what we want? What sort of world are we creating for our children?

Nowadays, children are often confused and overwhelmed by a society that has little real time for them. They are used as consumer fodder by manufacturers and advertisers, and so learn to demand goods which parents often cannot afford. They are often seen as an encumbrance when parents find themselves torn between childcare and work. Better provision of childcare by employers and the government, coupled with a more holistic approach by parents, is vital. Parents can at least look towards ways of meeting the problem halfway, by considering a jobshare, or finding a job nearer to home. We must learn to value the needs of our children, not just the material ones, but emotional and spiritual as well.

Collectively, we need to make our relationships with others greener. Is it fair for us to have clothes that have cost another worker their life because of pesticide poisoning? Products and goods sold to us in this country often use cheap labour from the Third World. Seventy five million children are subjected to appalling working conditions according to the United Nations, and millions more are denied medical and educational facilities while they work, often to pay off debts incurred by their parents. Manufacturing of some products causes environmental and health problems, while the mining and production of others can cause major pollution of the local area. Some families in poorer countries have stopped growing food for themselves while they grow exotic and luxury foods for us. In Ethiopia, during the worst part of the famine in 1984, the country had a bumper peanut crop.

Environmental groups internationally have helped smaller pressure groups in countries like Brazil and Malaysia to empower themselves and others. Supporting independent and cooperative businesses, being aware of the involvement of major companies and multinationals in these countries, and becoming less consumer orientated, all go towards helping individuals understand their role in holistic planet management.

Can our workplaces become greener in orientation? The answer is yes. There are many working groups worldwide where this is happening. Work should provide more for individuals than just a pay cheque at the end of the month! The process through which members of society learn and contribute their skills should be an evolutionary one. It is a depressing fact that most people do not enjoy their work at present.

However, training is now becoming more important for both businesses and workers, and new technology has taken some of the drudgery out of some types of mundane and repetitive work. Cooperative working, where management and workers have a say in the effectiveness of businesses, has increased productivity and workers' self-esteem. Often the people carrying out the tasks are best placed to see some of the problems, and greener businesses have proved that holistic management styles have produced more than just higher sales figures. Worthwhile, valued employment may seem like utopia to many, but it will become a necessity for those who wish to survive in the future.

Many of the ideas we've mentioned so far could be taught in schools. Educational facilities should be teaching more than just examination curricula for academic and vocational subjects. Our children need to learn skills which will enhance their lives and encourage a more holistic and developed sense of community and accomplishment. Learning should be playful and fun, and could so easily include confidence-building and assertiveness training; cooking for health; DIY skills; mending and sewing; group therapy and discussion; community involvement; stress management and other similar life-skills.

It is also important that we do not simply consider the rights of humans. All life forms on this planet have the right to live a natural life. How we interact with the natural environment is very important. Green issues first became popularized through public awareness of how our modern lifestyle was causing human illness and even death. Considerations of diet and exercise then developed into increased awareness of our surrounding environment, and we are now very conscious of global issues such as rainforest protection and global warming.

These issues have reawakened an awareness of our immediate environment – the home – and how it is a microcosm of the living earth. Problems we find in the home reflect problems found in the general environment. As consumers we *do* have control over the products we buy and bring into our home. As consumers we *can* dictate to manufacturers what we are prepared to buy. The strength of this 'consumer power' can be seen in the way that supermarkets have responded to the green challenge.

One way to protest about issues concerning you is to write a letter! For instance, if you want to influence the companies who are killing dolphins while fishing for tuna, it is not really sufficient just to stop buying tins of tuna – although this will make some difference. But the companies will not know *why* you aren't buying their product. The best thing to do is to write to them and list your concerns, always asking for a reply. This can be effective: for instance, 2,000 people wrote to the chair of Midland Bank regarding the bank's policy in South America; the letters prompted the chairman himself to visit the area, making a radical change in policy possible.

Large companies have a mandate to their shareholders to pay dividends on their shares. Because of this, individuals have responded very strongly to the environmental challenge. Some large companies, such as 3M who manufacture office equipment and stationery, have said that by the end of this century they want all their manufacturing process to be environmentally benign. They are also moving away from centralization, with their average plant having less than 150 workers, and ideas for new products coming from within the workforce.

Something of this approach can be used in our homes. For example:

1. Does everyone in the house agree with making the home green? If not, start discussions so that all sides of the argument are heard. Discussion of green issues with others helps everyone understand them better.

2. Plan how ideas can be introduced. You cannot do everything at once. What do you feel is the most important thing to start with?

3. How do these ideas fit in with the activities of the house? For example, are you out most weekends? And if so, are

there thermostats on the heating which you could alter to account for this?

4. If you are out most of the time, do you have the time to make your house green? Will you have to alter your schedules to do so?

5. How much energy does your house use? What methods could be introduced to reduce this? Can you invest in the most efficient energy-saving devices, such as solar power? Can you utilize sustainable energy sources whenever possible?

6. What materials are there in the house? Can these be sustained indefinitely without harming the environment? Time needs to be put aside to keep up with the latest information.

7. What happens to your household waste? Is it recycled? Do you have different bins for different types of waste? How can you reduce waste?

8. Is the house designed to prevent accidents? Does someone have knowledge of first aid in case of emergencies?

9. Do you have a household policy of what products you buy? Are they checked for chemicals and pesticides before buying?

10. What happens to produce surplus to needs, such as clothes that no longer fit or furniture no longer needed?

11. Where is your money invested? Is it put into funds that are ethically and environmentally managed?

12. Are the long-term plans for the house in keeping with environmental ideas?

13. Are there times when the house can be quiet and our senses can have a rest?

We are now under such pressure to consume so that industry can survive that we wonder how much longer it will be before all the earth's resources run out. As individuals, we have to start working for change. When others see that we are successful they will join in with the work.

Local activities are the best, as the results are the easiest to see quickly. Recycling is often a starting point in the change-over to a greener lifestyle for many people. Recycling schemes are becoming increasingly common, and it is easier than ever before to recycle the 6 billion cans, 11 billion bottles and tonnes of paper that we use every year. However, we really must ask ourselves why we produce so much waste in the first instance. The average person throws away 2 tonnes of domestic waste every year, much of which is unnecessary packaging to make the products look better than they actually are. We should be putting pressure on manufacturers, retailers and governments to curb this wasteful practice. As individuals we should stop buying overpackaged goods, and continue to use recycling facilities when necessary.

Be aware of what is happening in your local environment and in the area local to your house. Watch out for people dumping rubbish in open areas. The more that we stand up and complain when these things happen, the more likely that others will take up the campaign as well. Fly-tipping causes unsightly mess. It can be anything from matresses in the streets to toxic waste dumped on public tips; all causes a myriad of environmental problems. Watch out also for the cutting down of trees. Complain to the local council if you suspect that this is happening. Read your local paper for any new planning application which might cause destruction to the environment, especially where the change of use of the land is being requested.

Politics is another area in which we can make changes. Political activism means more than just having a vote every few years at a general election. We can be politically aware in our local environment too. Decisions at local and national levels are made by a few people largely because others let them. We *can* have a say! If politicians ignore your worries and concerns then you can always vote them out of office or even stand for election yourself. Each major political group has made policies on major areas covered in this book, including the environment. You can write to them asking them what they are, before committing yourself. Don't just go by what your daily newspaper says – they are often biased.

Politics includes ensuring adequate access to information about the products that we buy, so that the consumer can make an informed choice. We must lobby manufacturers and governments to make information available: we need adequate nutritional information on food products; information about energy efficiency on electrical products should be clearly stated; we should demand to know about the country of origin of products so that we can make informed decisions on ethical grounds. We should be allowed access to information on the polluting effects of certain manufacturing industries. If we are not given the right of access to information we are powerless to initiate changes for the good of our environment.

Going to the AGMs of companies you may have shares in is another way of influencing the policy-makers. Read the annual report, and go to the AGM to ask questions about anything you do not understand. You can also ask about company policy at these meetings. The more people go to these meeting the more the directors of the companies will take their investors' views seriously.

Doing all of these things represents a major life-change for many people. Some things are more difficult to achieve than others. But remember, each of us can start with small things, working our way gradually towards a greener, safer, healthier life.

Remember that Green Living is more about an attitude to life rather than about buying something different. Part of that attitude means having fun as well as being serious. Good luck!

1 Building and Design

Introduction

Creating a building means to bring inner pictures into the light of consciousness and make them real.

M. Schimmelschmidt

Our homes have always been the symbol of safety and nurturing; the places where we grow and learn, and where we have privacy and protection from the outside world. While we seem to be very concerned about the effect of our lifestyle on the environment *outside* our homes, few of us are thinking about the effect our environment *inside* the home has on our health and well-being.

In many countries, the building and design of the home has been left to the occupant and has been designed completely for their needs. Only in industrial, Western, so-called 'developed' countries have we relied heavily on architects and project planners. Post-war housing has come under severe criticism, especially the increase of high-rise flats and the over-use of concrete which creates a cold and depressing atmosphere. Sociologists and anthropologists have been studying the emotional, physical and – more recently – the environmental effects that vandalism, poverty, and isolation cause.

A pioneer of green architectural design was Rudolf Steiner. His Goetheanum centre in Switzerland was and is an inspiration to many architects. He was one of many artists and architects who, at the beginning of the twentieth century, looked to nature for inspiration. Mackintosh and Gaudi were similarly inspired, and they believed that nature herself contained solutions to problems of style. Steiner started to work first with colour, by painting walls with a glazing technique which gave the impression of luminous transparent colour. He advocated transparent colours above all, as he felt they reflected the cosmic play in nature. These paints were the first of the natural paints that we now have in the organic paint range. A German architect, Tant, applied the idea to exterior colour, and a wide range of ideas flowed from there.

Steiner's ideas on indoor wall painting were soon to blossom into a new architectural style. His aim was to make every straight line lead 'over into the curve, balance is sought everywhere, the endeavour is made to melt what has become frozen, so that it may flow again, rest is everywhere created within movement and is again caused to move' (Steiner, 1918).

His buildings were not intended to cut the occupant off from the world. He stated:

> *if we carry this spacial system of lines and forces, constantly active within us out into the world, and if we organise matter according to this system, then architecture arises. An architecture consists in projecting into space outside ourselves, the laws of the human body.*

Steiner's building designs, like those who have since followed his ideas, have tried to reflect our part on the earth, with the earth, rather than a construction on the earth. His buildings 'arise from the earth and reach for the sky'. They are shaped like the trees and the plants and bend with the natural landscape around them. They are made only with locally-found materials, and so reflect the landscape rather than standing out against it. By most standards, Steiner's work is remarkable and unusual. Sadly, most of today's architecture lacks any sense of this collusion with the earth itself; the most important consideration seems to be price.

However, some recent initiatives are more encouraging. New community housing has encouraged people to get involved with planning and design at an early stage. Housing co-operatives, self-build associations and home-building projects all foster an increased participation and constructive relationship between needs and wants, and offer a new chance for individuals and families to create the sort of environment important for them.

Building and Design

The combined marketing and legal strength of the building industry has almost convinced us that we cannot build our homes ourselves. But there are now a growing number of individuals and groups who have built their own homes and survived! Courses are available, books explain in detail how to do it, and there are design and management courses for associations wanting to help groups do it together.

Over the last hundred years the design of our homes hasn't changed too radically. We still build to more or less the same standard design that we are accustomed to, with bedrooms, bathrooms, general living areas and so forth. Our living standard however, has improved beyond measure in some areas. Almost everyone has a flush toilet, access to electricity or gas heating, lighting, and running water. The quality of our lives in these homes may have progressed in this sense, but increasing awareness about the levels of indoor pollutants has caused us to stop and think about some of the ways in which this 'progress' may be affecting us.

Some doctors have begun to study the many symptoms of individuals suffering from the environmental effects of their homes. Some people are developing 'environmental sensitivity' to buildings, to the chemicals used in the treatment of building materials, or to the design of an 'airtight' building leaving them without clean air to breathe. This sensitivity can manifest itself in many ways. Doctors have noticed effects both on the skin and on the major organ systems. With toxic or chemical overload the symptoms change according to the amount and type of chemical that triggers the attack. Symptoms can include a general feeling of malaise and fatigue, temperature sensitivity, unexplained spots on the skin, swelling of hands and feet, impaired judgement and memory loss, and even an intolerance to medicines.

Current building design and technology may be the cause of some of the problems. Air pollution inside our homes has been trebled by the massive jump in the numbers and types of synthetic chemicals used in building materials. An estimated 5,000 different types of chemicals are used in modern homes. If concrete tiles are used on the roof there will be a high resistance to atmospheric irradiation, and cavity walls filled with dense sand and cement can stifle breathing (or diffusion) through the walls. Steel lining can create an unhealthy magnetic field, and synthetic flooring prevents the natural earthing of our bodies' electric charges, as well as contributing to the indoor pollutants from chemicals used in tiles and vinyl floor coverings.

But it is not all gloomy. Recent work has established a wealth of new ideas for green designers who want to build a healthy environment with the minimum impact in terms of pollution and energy use. A healthy home environment is called a 'bioclimate' by architects of the German 'building biology' tradition. The three most important aspects are the air quality, the electroclimate and the microclimate. Building biology looks at these aspects and how design and construction techniques can avoid indoor pollution and enhance the 'bioclimate'. For example, a surprising amount of fresh air can diffuse, or 'breathe' through a solid wall, thus helping to replace stale or polluted air. To do this adequately the design needs to encompass an unobstructed cavity, a good proportion of joints, a porous inner cavity and a plaster finish of gypsum with paper that doesn't seal the wall.

To design a home from scratch is probably an impossible dream for many of us, but at some time major parts of our home may need to be renovated. This might then be the perfect time for you to take a good look at the structure of your home and see what you can achieve.

Basic principles of environmental design

1. Houses should be re-used or adapted rather than demolished.

2. Siting is crucial. Questions worth asking include: what is its solar orientation? What is the local area like? What are local weather conditions like? Which local architectural materials are available?

3. Does it provide a harmonious environment for those who are going to live in it? Will it be comfortable? Will it be a statement of those that live in it or is it a statement of the designer or planner?

Building and Design

> 4. Does it use local materials in its building or conversion, rather than materials from further afield?
>
> 5. Is it energy efficient, as well as making use of recycling waste?
>
> 6. Do the interior materials used have a long life? Natural materials, though more expensive in the short term, will have a much longer life.
>
> 7. Natural ventilation and natural light should be a feature of its design.
>
> 8. All parts and surfaces should be built to last at least one hundred years with a full usable life.
>
> 9. All parts of the home should be easy to maintain and repair.
>
> *Source*: Avril Fox and Robin Murrell, *Green Design*, (Longman 1989).

Environmental houses have been built in several cities in Britain. They offer a great example of design in progress and many have been built specifically as show houses so that the public can visit. Houses such as the Leicester Green House, the Survivor House in Powys, the Green House Project in East London, the Underground House in Gloucestershire and the houses-building project at Findhorn, Scotland can all be visited. See page 199 for addresses.

Directory

Take green living to the fullest possible extent and build your own house. You don't have to live in a teepee either! Many self-built homes are luxurious and comfortable. They are of course built with the owner in mind, so they will be designer-made and much more useful. To build your own completely green house would mean lots of research and hard work, but the rewards will be immense. Others have succeeded and the idea could become a hobby in itself; it is becoming quite common in Europe and North America.

The ideal green house would contain the following features:

- a high level of insulation;
- solar water heating;
- passive solar heating including a conservatory;
- heat recovery from ventilated air and water waste;
- collection of rainwater for toilets;
- recycling of water for garden use;
- building materials chosen for their low environmental impact (i.e. using few hardwoods, extensive use of materials requiring little energy to produce, and if possible, renewable);
- construction materials which are durable, easily repaired and maintained, and easily dismantled at the end of their life.

Here, in brief, are some things to consider both in building a new house and in doing renovations.

Dampness
Dampness can be a danger to health, as can the spores of fungus that grow in damp conditions. Ensure that dampness is not being caused by any trees or plants that are close to the house. Ensure that all drainage is away from the house. Around foundations there should be a porous border to allow for drainage. Good ventilation will also reduce the risk.

Electricity
Even though we can produce our own electricity by solar power, electric cables in the house and under the floorboards still produce electromagnetism. Try to make sure the cables do not go right around the room. In a bedroom try to have sockets on one wall only, leaving the rest of the room free from electrostatic interference.

Energy
Get an information pack from the energy efficiency office on how energy saving can be made in the home. Draught strip all the doors to save on heat. Place shelves above radiators to improve the radiators' efficiency. Also, you can stick tin foil behind the radiator so that heat is reflected into the room

rather than the wall. See the chapter on energy for more information on energy efficiency measures.

Floor Insulation
The floor of a house will not lose as much heat as the walls or the roof, but if you are building your house from scratch, floor insulation should be a consideration. Ground floors are usually made of concrete laid down on the foundations or from wooden timbers attached to joists with a ventilated space between. Heat loss occurs through the suspended timber flooring because of the draughts this creates. The flooring must be well-ventilated to prevent the build-up of moisture, which encourages dry rot. The most thorough solution is to lift the boards and insulate between the joists. Concrete floors cannot be lifted of course, but you can further insulate them with layers of tongued and grooved flooring-grade board obtainable from builders merchants.

Garden Design
Garden buffers of trees and shrubs can help the house blend in to the landscape, provide privacy, and reduce the noise from the house. Fences and screens well placed can also reduce the noise pollution.

Green Architecture
Green architecture and green architects look at the way buildings are designed and built, constantly keeping in mind the ideas of the building biologists. The main concerns of green architecture are: the materials used to build the houses; the effect that the building may have on the occupants; ensuring minimum environmental impact. Minimum environmental impact is achieved through the use of solar energy and natural products, so that the energy consumed to make and live in the house is reduced.

The chemicals used in most building materials are toxic as they contain chemicals like CFCs and formaldehyde. When the buildings are finished they continue to give off tiny amounts of chemicals into the home. Green design, by contrast, creates an atmosphere in which the air is fresh and clean; this is done by having a house which breathes. Houses are built with porous materials like lime mortar for the bricks, and gypsum plaster. Many houses built before the war have this sort of structure anyway. This does not mean that there is increased heat loss from the building, as the air only moves very slowly through these surfaces. Some architects say that we have an emotional response to natural products which we do not have to plastics and metals. Those that favour the emotional response say that by touching and smelling the natural woods and stones in our houses we have direct contact with nature. Metals are explained more fully on page 53. They are not considered a 'natural' house material as they would have to be heavily processed. Metals and plastics divorce us from this natural contact.

Plumbing

Building and designing plumbing which saves as much water as possible is an important consideration in any environmentally conscious home. A toilet which is plumbed in with a dual flush mechanism will save nearly one hundred litres of water on an average day.

Roofs

Roofing Treatments

You will want to protect the health of wildlife as well as your own health. The Nature Conservancy Council suggest using zinc naphthenate if you have to use any wood preservative treatment in your roof, as this is relatively low in toxicity and is safer for any bats that may be in your roofspace. (However, these are still not completely harmless and need to be used with great care.)

Grass Roofs

If you really have no space for a lawn why not put one on the roof? These are very popular with the biological building schools, but of course are only relevant if you have a flattish, strong surface. The earth is a very good insulator of heat, and you could save extra energy.

Solar Roofs

The roof in the United Kingdom is ideally designed to take a solar panel, a very effective way of heating the water. (Solar immersion heaters can also be used to heat swimming pools!) Solar heating offers a non-polluting saving in energy costs. If the insulation is well-designed and soundly installed, the saving achieved will be well worth the investment.

Turn the roof into a conservatory and grow trees and shrubs. The heat caught here can be convected into other rooms and also used to heat water.

Roofing
Galvanized steel or painted aluminium is sometimes used on roofing; it could have oily residues, which have to be removed. Plastic, rubber, asphalt, shingles and wood shingles have all been found to release contaminants, particularly when heated. Slate, although quarried, has a long life and can be reused many times.

Room Placement and Lighting
When moving into a house or flat, think carefully about the placement of the rooms; which rooms get the sun and which do not. As a general rule of thumb, those that get the morning or afternoon sun are good for meeting places such as kitchens and living rooms. North facing are good for working in if the office is at home, as long as they do not have large windows, as they will be difficult to heat in the winter.

The design and placement of lighting in the room is very important. Natural light is of the utmost importance. Remember to have curtains pulled well back off the windows so they do not block any light. Uplighters give good light which does not cause glare.

Seasonal Design
As the year passes by the seasons change, and the design should reflect these changes by providing focal points for the winter, and sunny spaces full of fresh air for the summer. For example, fires for the winter months, and conservatories for the summer.

Shutters
In the winter more heat is lost from the house through a single glazed south-facing window than is gained to the house from the sun shining through the glass. If the same glass were double-glazed and fitted with insulated shutters which were closed at night, the window would act as a solar collector, and energy could be gained over the winter months.

The Sick House

Houses can be filled with pollution from the chemicals used in the production of the construction materials. Pollutants come from the energy used in the house and may even come from the ground that it is standing on. All houses will contain some pollution. What we need to do is try to minimize it.

Solar Energy

The amount of solar energy available in Great Britain is much greater than we imagine. On a sunny day we are aware of the sun's heat, yet even on a cloudy day there is enough scattered and diffuse radiation to be useful. The amount of solar energy falling on the roofs of our houses is many times what it is required to provide us with all its hot water and heating. The average amount of solar radiation received in a year in this country is 60 per cent of that received at the equator. This radiation is the equivalent to the output of 1,000 power stations. If every house was equipped to collect solar energy there could be a saving of 40 million mega watt hours. The problem is that at present, the technology is available at a high cost. It is generally worth considering if you are renovating or building your own home from scratch.

The advent of the space satellite was directly responsible for the further development of materials that can convert the sun's energy into electricity. Reliability rather than cheapness was the requirement in the early days. The expensive silicon crystal that was first used has been replaced by amorphous silicon. Other chemicals are also being researched, some of which can be deposited on ribbon substrates that allow square cutting. We can now buy solar clocks, watches and calculators as well as security lighting, solar door bells and garden lights.

Solar Water

Domestic water systems need additional heating by conventional means during the winter months, yet for the solar energy houses built at Milton Keynes, the energy costs can be substantially reduced during the cold winter months. To have all our heating for the house from solar at this time requires a very heavy expenditure for the equipment. Solar energy offers a non-polluting source of energy. If the installation is well designed and soundly installed the saving achieved will be well worth the investment. In northern latitudes, solar water heating panels will pay for themselves within five to fifteen years.

Solar Heat Traps
By far the best way to make use of solar heating is to make our houses into solar heat traps. Passive solar heating can be incorporated into existing houses by building conservatories on the south facing side of the house and painting south walls a dark colour. The Trombe Wall is one design in which air is heated up by the sun shining on a black glass covered wall. The warm air is circulated through slits in the wall and is another example of passive solar heating. Passive solar heating can be used with any heat-absorbing surface which can be incorporated into a wall.

These south facing walls need to be unshaded with wide frontages. Living areas, glazing and conservatories should also be on the south side. The north-side rooms can be used less frequently and have smaller windows. Besides providing bright and light rooms 10 per cent saving in fuel bills can be made by following passive solar design.

Traffic flow
In households good design of the most travelled parts of the house and how the traffic flows is important for reducing irritation from intrusion and interruption.

Ventilation
Good ventilation and movement of air through the home is essential. Air should be able to move from one area of the house to another without going straight through, so that the house can be filled with moving air. This is important so that pollutants do not build up in rooms.

Wind Power
Unfortunately domestic wind-power generators are rarely able to provide enough energy for heat or power, but they can be used to charge batteries and to provide lighting. The biggest problem is that it is very difficult to harness and store the massive energy from wind power over long periods.

Wood preservatives

Wood preservatives are used to kill organisms in timber and so are always toxic. Most can last for twenty years and the most commonly used are suspected carcinogens. Some still available in the UK have been banned abroad, like Dieldrin, and others which are responsible for large numbers of deaths like pentachlorophenol (PCP) which has been implicated in the deaths of over 1,000 people worldwide. Over 3 million tonnes have been chemically sprayed in Britain, mostly in roof spaces; the production, transportation and disposal of these chemicals have added to water and air pollution. In most cases you won't need to do anything with your wood, especially if it has good ventilation and if airbricks are used. Care and maintenance will avoid dry rot and infestations. Only the exposed timbers in roofs may need extra attention and treatment.

2 Power and Energy

Introduction

Taking dwindling resources and the greenhouse effect into account, the answer must be energy efficiency and conservation. This is one area where it is easy to maintain standards of living while consuming less. Solving the equation any other way would be madness.

Amory Lovins

We all need energy in our homes, to keep us warm, to allow us to cook, and to run equipment. Most of the energy we use comes from electricity or fuels like gas, with a smaller percentage coming from alternatives like oil, propane and even solar energy.

Our use of energy is responsible for a large number of environmental problems. The pollution caused by fossil-fuel use ranges from acid rain (created when sulphur dioxide, nitrogen oxide and nitrogen dioxide gases mix with water in the atmosphere), to carbon dioxide emissions (leading to the greenhouse effect), and even oil slicks (which kill large numbers of wildlife and endanger ecosystems). This large scale pollution is directly connected to our use of energy at home. Every time we turn on a kettle or switch on a radiator, various gases pollute the atmosphere. Each time we use an electric immersion heater for a shower or bath, we indirectly produce nitrogen oxides, carbon dioxides and sulphur. In the United Kingdom each individual is responsible for using the equivalent in energy terms of six tonnes of coal each year. In the United States, consumers are responsible for using the equivalent of over twelve tonnes each of coal. We each use up to fifty times as much energy as someone living in a developing country.

We rely on energy sources to maintain our lifestyle. But the fact is that most of these have come from finite resources. The sun produces most of these fuels over millions of years: coal, oil and gas are all various forms of stored solar energy laid down in rock formations. Over the last two hundred years we have been using up these fossil fuels, usually without thought for the future. Scientists argue that we have several hundred years' worth of coal left, only 60 years of gas and 40–50 years of oil at the current rate of use. There are uncertainties in these calculations, research is finding new sources of fuel, and of course some fuel companies are unwilling to give a clear idea of how much is left. But the overriding evidence is clear: we are running out of our fossil fuel resources, and, in the future, costs will increase.

Fuel Sources

Gas is the most energy-efficient way to heat space and water for most consumers in the United Kingdom at present. There is less pollution in the form of sulphur, nitrogen oxides and carbon dioxide from the gas itself, and burning gas in efficient boilers wastes much less energy than conventional power stations generating electricity. However, gas is by no means a completely clean fuel, its use releases methane, a powerful greenhouse gas, into the atmosphere.

Thirty-five years ago scientists came up with what they thought was the perfect alternative to fossil fuels: *nuclear power*. But nuclear power, using uranium for its fuel, has been heavily criticized by environmentalists and energy experts. The safety record of various nuclear power stations has come into question, especially after the tragic nuclear accident at Chernobyl, where about 100,000 people are expected to develop some form of cancer as a result of the core meltdown of the reactor. It may take between ten and forty years for the sickness and deaths to occur and the accident has already cost £30 billion. Other accidents, like the one at Three Mile Island in the United States have fuelled speculation about the long-term safety of the nuclear power industry, and many governments have completely abandoned nuclear energy. The costs of nuclear electricity make it uncompetitive in countries such as the United Kingdom and United States.

At present, nuclear power represents only 3 per cent of the

world's commercial energy. One of the major reasons for opposing nuclear power is the concern about the large amounts of nuclear waste that are created. The waste is reprocessed in the United Kingdom at a plant in Sellafield, Cumbria, where workers were recently told that they have a one in three hundred chance of giving their future offspring leukaemia. Several families are now trying to take legal action against the British nuclear industry. So far, waste has been stored in large tanks on site and will be active on our planet for several thousand years. Plutonium, one waste by-product, is used in the manufacture of nuclear weapons. It will take a staggering 24,000 years to reduce its level of radioactivity by 50 per cent. This sort of risk for the public leads many people to conclude that nuclear power does not offer the answer to our energy crisis.

The continued use of *fossil fuels* and the subsequent production of carbon dioxide increases the likelihood of global warming. Over 50 per cent of the greenhouse gases responsible for global warming are caused by our use of energy. Carbon dioxide comes directly from burning fossil fuels such as coal and oil. Nitrogen oxide, another greenhouse gas, is also produced by burning these fuels, and wood. Some of the greenhouse gases are naturally occurring, but the energy we use is increasing the levels of these gases so quickly that natural processes cannot absorb them.

Problems on the Increase

As the world's climate changes we could expect to see *coastal flooding* in low-lying countries such as the Netherlands and Bangladesh. We can expect more freak weather patterns like tornadoes, floods and hurricanes. Scientists have warned of massive plagues of insects, and our crops might not be able to grow when we expect them to, so supplies might become unpredictable and erratic. Glaciers would melt as the poles become hotter, leading to flooding and a raising of sea levels, and ecosystems will be disrupted, causing problems for the lifestyle and habitat of many animal and plant species. Many will become extinct and others will have to move in order to survive.

Acid rain is caused by gaseous emissions from power stations and cars. The more electricity the power stations have to produce for us, the more pollution they create. Rivers, lakes and forests are at risk in North America and Europe from acid rain. Over 4,000 lakes in Sweden are almost dead, fish die and other wildlife in the food chain are affected. Forests are said to be dying due to acid rain. Even buildings are affected, as the sulphuric acid dissolves the stones and concrete. Many of the greatest buildings and monuments in the world have been affected. In Greece, the Parthenon on the Acropolis, which was built around 447 BC, has suffered more damage from pollution in the last thirty years than in all its previous history.

The Way Forward
Believe it or not, we don't need to use all this energy. We can live with the same level of comfort and service, but using much less energy. Insulation of the home is one major – and simple – way to reduce our energy consumption. We may as well have a huge hole in the nearest outside wall of many homes at present, such is the lack of care about energy consumption. We could reduce our current consumption by at least 30 per cent by making some fairly straightforward changes.

For most houses, the biggest energy bill – representing 70 per cent of the energy used – is for heating space to make us comfortable, and heating water for cleaning and cooking. Lighting and appliances take up the rest. The average United Kingdom household is responsible for producing 17 tonnes of carbon dioxide each year. It is possible to reduce the amount of carbon dioxide that we produce by using less energy and by being more efficient with the energy that we have.

Of course, saving energy isn't only good because it saves on pollution; it also saves people money. Who can really afford to heat a hole in the wall? Most standard energy-saving and efficiency measures can have a very short pay-back period. That means that the original cost is met by the savings that you will get from it in a short time. A thick lagging on a water tank can take as little as a few months to pay back the original cost; after that, you are literally making money! A few pounds spent on draught-stripping around windows, doors and skirting boards will warm up your home considerably. Then, of course, you can turn down the heat and save even more cash. Even a letter-box can be insulated to stop howling draughts in the hallway.

An energy-efficient home has other benefits too. For example, energy efficiency cuts down on condensation; condensation causes the deterioration of wallpaper or paint and woodwork. Dirt and grime also attract themselves to cool surfaces; a well-insulated home can stay cleaner longer.

One important reason for taking care of our energy consumption is to provide cheaper, more efficient means of heating for the elderly. Many people can no longer afford to keep the heating on all day and have died from hypothermia and lung or bronchial ailments. Good energy efficiency will stop them spending so much money and keep their homes warmer.

Simple Steps for Energy Efficiency and Saving Money

1. Do an energy audit of your home: check out the best and worst spots for draughts, for example. A simple leaflet is available from the Energy Efficiency Office.

2. Add up the cost of your heating and lighting, you may be surprised at how much it costs. Each home could save an average of £100 a year with basic efficiency measures.

3. If you are an OAP or on a low-income, contact Neighbourhood Energy Action who operate several hundred local insulation projects.

4. Insulate your hot water tank–it will cost less than £10.

5. Ask for energy-efficient lights or draught-proofing materials for Christmas or birthdays–it could be putting money in your pocket!

6. Close up rooms that you don't use; don't heat each room unless you use them each day. Use thermostatic radiator valves (TRVs) to regulate the temperature in each room.

7. Use cold water more often for rinsing and washing.

> 8. Check all your appliances before you buy them: do you really need them? Can you borrow them from friends? Are they the most energy efficient available? How much will they actually cost to run?
>
> 9. Offer to help friends to insulate their homes, lag water tanks, and so on, especially the elderly or people living on their own.

Directory

Air Conditioning
Though our temperate climate hardly justifies it, sales of domestic air conditioning (AC) systems have increased considerably in recent years. AC systems use as much electricity as a fridge freezer, and CFC's are used in their cooling systems. Keep cool by wearing light, loose clothing in the summer and by drinking lots of fluids. During warm weather open your windows during the night and close them during the day to trap cooler air inside. Shade windows in bright sunlight with blinds, or go continental and fit shutters.

Appliances
There is a variation of 300 per cent in the amount of energy appliances such as fridges or washing machines use, and in the way that people use them. To cut down on the energy consumption of your household appliances, you should purchase the most energy-efficient models available, and follow the advice offered in the equipment chapter, where individual items of equipment are listed.

Baths
Twenty pence out of every pound spent on energy is used for heating water, so it is worth taking a closer look at the amount of hot water that you use. It takes an average of 20 gallons of water to fill a bath, whereas you could take three showers for the same amount of water. By changing to daily showers from daily baths, you could save 92 gallons of water a week, saving money and energy.

Batteries

More than 400 million batteries are used and thrown away in the United Kingdom every year. The manufacture of the battery will have cost 90 times the amount of energy that the battery will actually produce during its lifetime. They contribute to toxic waste and you should try not to dispose of them in your domestic refuse. Check if your local authority has a battery disposal point. Less toxic alternatives are now available: a battery recharger (preferably solar powered) will cost about £15, and, with a good stock of reusable batteries, would be a far better choice.

Boilers

Your central heating boiler should last you for at least fifteen years. If you are buying a new one you will discover that they are now considerably smaller and more energy efficient. If you have gas or oil, it would be worth considering a condensing boiler, which will use 85 per cent of possible heat in the gas burnt by the boiler, compared to 60–75 per cent of older boilers. If you wish to use solid fuel, a coalflow boiler will run for a week without refuelling. Whatever sort of boiler you decide to invest in, try to buy the most energy efficient that you can. It may cost more initially, but you will soon reap the rewards in terms of smaller fuel bills. Remember to keep your boiler well serviced for maximum efficiency. Good insulation of your house will also cut down on the demands made on your boiler.

Cavity Walls

As much as a third of all our household heat is lost through our outside walls. Most houses built since 1930 have cavity walls with a 50mm (2 inch) gap between them. If you fill this gap with insulating materials you can save up to £100 a year on your heating costs. Once insulated this way, it is estimated that your home will produce 100 tonnes less CO^2 during its lifetime. An experienced contractor will charge between £350 and £600, depending on the size of the house, so you will be looking at a 4–6 year pay-back time. There are several types of material available. Try to avoid UF foam, which gives off formaldehyde fumes (they can cause allergic reactions) and do not use polystyrene materials that use CFC's.

Ceilings

Polystyrene ceiling tiles, while cheap and effective, are a severe fire risk, and CFCs are currently used to manufacture them. Foil-backed plasterboard reflects heat back into the room, and when it is combined with suitable loft insulation, can minimize heat loss. In older buildings with high ceilings, suspended ceilings made from rockwool or similar materials will again reduce heat loss. In open plan houses, or ones with large stair wells, a suitable fan can recirculate the warm air.

Central Heating

Although 77 per cent of our homes are centrally heated, many of these systems are inefficient, costing us money and contributing to global warming. We should therefore ensure that our heating systems operate as efficiently as possible. This means that boilers should be serviced annually (older oil boilers twice annually) and other components such as the circulating pump and safety vent pipes must be kept in good condition either by the keen DIY expert, or by a professional.

Once we have observed that the system is working as well as possible, we should look at other ways of reducing our fuel consumption. By turning our heating down by 1°C, we could save 8 per cent on our fuel bill (The Centre for Alternative Technology recommend a maximum temperature of 65°F for our homes.) Fit a thermostat with a time programmer and you could save up to a further 20 per cent on your fuel bill. You should also fit individual thermostatic radiator valves to individual radiators in each room. (See entry on radiators.)

Chimneys

If your chimney or flue is unused, it would be sensible to seal it off to avoid substantial heat loss. You will need to block up the fireplace openings and fit a ventilated cap on the top of the stack, and you will need a ventilation grille in the chimney breast or flue to prevent the buildup of condensation. A chimney repair specialist (listed in Yellow Pages) would be able to advise you on this matter.

Cooking

As we mentioned in Chapter 1, if buying a new cooker, look for special energy saving features, such as dual elements on the hob. Try to limit your cooking as much as possible. Buy a pressure cooker, make a haybox (see pages 38–9) or discover the delights of uncooked foods.

Curtains
Glass windows allow rapid heat loss from rooms, so well-fitted curtains can provide a necessary layer of insulation, especially in the evenings and overnight. Try to buy lined curtains (the heavier the fabric the better), or insulated roller blinds. For maximum effect, the curtains should be fitted inside the window opening with the curtain ends stopping on the sill to prevent down draughts. An even better alternative would be ceiling-to-floor and wall-to-wall curtains which would help to prevent loss of heat through the whole wall. A further consideration for preventing heat loss from windows is outside shutters. They keep the cool in during summer and cold out during the winter.

Doors
A great deal of heat is lost because of badly-insulated doors. The problem is easily remedied by simple draught-proofing measures. All sorts of draught excluders are available, ranging from a simple adhesive strip to a sophisticated threshold sealer. None should cost more than a few pounds, and you will recoup the cost in a few weeks through saved fuel expenditure. Remember your letterbox and keyholes too. An extra flap inside the door will prevent any persistent draughts. Look out for soft toy draught excluders at craft fairs and bazaars. They really do prevent the cold from creeping under doors by forming a barrier.

Double Glazing
Unless you are already intending to replace windows, think carefully before installing double glazing. As a solution to blocking unwanted noise, it is effective. In terms of energy, however, it is not necessarily a cost-effective form of insulation; you may find that you never break even on the cost. Secondary double glazing is a cheaper alternative, or a simple form of double glazing can be achieved by tacking plastic sheeting around your windows. It will only cost a few pounds to do, and will cut heat loss. The added advantage is that it can be taken down in the summer!

Floors

Doors and windows are not the only source of draughts – floors can create similar problems. Gaps in the floorboards and between floor and skirting boards can be sealed with a sealant gum. A suspended timber floor can be insulated for less than £150 with loft insulation materials, as long as you are prepared to take up the floorboards to do it. If you have solid concrete floors, a thick carpet with ample underlay will offer good insulation.

Gas

Gas is the most common fuel source for central heating in the United Kingdom. Around 76 per cent of existing systems and 80 per cent of new systems are powered by gas. Electric heating results in the emission of three to four times more CO_2 in heating a home than gas heating. When you examine the comparative cost for space heating per kilowatt hour (kwh), gas emerges as the most economical. An electric storage heater run on Economy 7 tariff will cost £2.75 per kwh, whereas a gas radiator using a condensing boiler will cost £1.60 per kwh. Gas is also cheaper for cooking and other household uses, and again, produces far less CO_2 to add to the greenhouse effect. So if you do have a choice, make it gas.

Grants

If you are receiving income support, family credit, housing benefit, or you are on a state pension, you may qualify for a grant of 90 per cent of the cost of insulating your home. Contact you local Social Services department, or the Welfare Rights Department at your local council for further details.

Haybox

A haybox is an easy way of slow-cooking your food, especially good for stews, soups and casseroles. The beauty of it is that it doesn't use any energy. You can buy commercially produced hayboxes, but it is far cheaper and much more fun to make your own.

Get a cardboard box that is large enough to fit a saucepan inside, with a space around it. Fill the box tightly with dry hay, or crumpled newspaper. Put the ingredients in a saucepan and boil for 10 minutes, then put the pan straight into the haybox and cover with even more straw. Close the flaps on top of the box and seal them with tape. Leave the food to cook for 6–8 hours, and you will have a tasty casserole or soup to eat at very little cost.

Heat Exchanger

A heat exchanger works by removing heat from one area and directing it to another area where it is needed. Thus hot air in an office or factory can be removed to heat water, or for a central heating system. Alternatively, a positive temperature externally in either the air or an adjacent lake or reservoir can be tapped and used in a house to provide heating and hot water with a heat pump. This could reduce present central heating requirements by 30 per cent in some circumstances. There is not yet a cheap system to install, however, and the pay-back period could take quite a few years. But, as the system becomes more common domestically, and therefore cheaper, there is no reason why we should not use heat from our washing machines, baths and cookers to complement our existing heating systems. Contact your local Electricity Board for further details, and the Electricity Council for a special leaflet.

Humidity

Central heating often causes our homes to be too dry, which can lead to sore throats and other minor ailments. It also increases the build up of static electricity, which produces too many positive ions and can lead to health problems, particularly for people with asthma or bronchitis. The simple solution is to open windows from time to time, and leave bowls of water around. You can buy humidifiers and ionizers from many electrical stockists, but try the simple remedies first.

Labelling

We should lobby our government to introduce legislation making it compulsory to label the energy consumption of electrical items. If the public were given this information, they could make informed decisions about the goods they were buying. Lobbying worked to give us better product information in the food industry: we are now given a comprehensive breakdown on nutritional information. Write to your MP and inform him/her about this matter, stating what you want him/her to do about it.

Lighting

If you replaced three of your ordinary 60w light bulbs with low energy compact fluorescent bulbs, you would save twenty pounds a year on your electricity bill. An 11-watt compact fluorescent can replace a 60w bulb in terms of light intensity. It will use 20 per cent of the power consumption, and it will last 8 times longer. The disadvantage of these bulbs is that they are expensive (between £13–£15) compared to incandescent (less than £1). But their energy savings and length of use will allow you to pay back these costs in a year or so. You could get together with friends and neighbours and buy a supply at wholesale cost to save money. They are now available at most electrical dealers and some department stores.

To replace all the bulbs in your home with compact fluorescent would be a costly exercise, so replace them gradually and in the interim use low wattage bulbs in other areas of the house where possible. The least expensive way of saving money and reducing the energy consumption of lights, is to turn them off when you don't need them. Make it a habit, and encourage the other members of the family to do the same.

Lofts

If your loft is not insulated you will lose up to a quarter of your heat through your roof. Most homes have less than the current standard of 150mm (6 inches) insulation in their lofts. Check inside your roof to see how thick the insulation is. Once insulated, loss will be reduced by 75 per cent. It is a fairly easy job to carry out yourself, and will cost you about £120 (£250 installed by a builder). It will also save you about £50–£100 a year, so the return makes the investment worthwhile. There are many choices of material available. For technical and environmental reasons avoid granules or beads that involve CFCs. A better choice would be cellulose fibre made from recycled newspaper.

While you are in the loft, don't forget to lag the cold tank and any pipework. That will avoid your pipes freezing in the winter.

Meters
Learn to read your electric and gas meters and bills so you know exactly how much fuel you are using and how much you are being charged for it. Try switching all your electric equipment off, then switch them on one by one, watching how fast your meter moves. That way you will really notice the difference when you have followed the advice we have given. Your local electricity and gas showrooms will have literature to help you to do this.

Porch
If you fit a small porch outside your house, it will act as an airlock and will reduce the exchange of cold air for warm air every time the front door is opened. The pay-back period may be a few years, depending on how elaborate you make your porch, but it will pay for itself, given time and give you a little extra space. Alternatively, fix a second door inside your house for the same effect.

Radiators
As mentioned in the entry on central heating, you can buy thermostatic radiator valves, which work by reducing the flow of hot water to the radiator when required room temperature is reached. You can fit these yourself, or a plumber will be able to advise you.

Another way of making your radiators work really well for you is to fix kitchen foil on the wall behind them (with the shiny side facing into the room). This reflects heat back into living space. If you have radiators under a window, don't let curtains hang over them, as it encourages heat which could be used in your room to escape through the window. Tuck the curtains behind them or place shelves above your radiators to direct warm air back into the room.

Solar Energy
Solar collection devices were being experimented with in Northern Europe as long ago as the seventeenth century. Solar energy is a vast resource that can be tapped for millions of years to come. Even in a country such as ours, with relatively few hours of direct sunlight, diffuse sunlight through cloud can be sufficient to meet space and water

heating needs in a domestic setting. There are three types of solar collector: active, passive and solar cell.

Active
Solar panels use the energy from the sun to heat water which circulates through a roof panel. The energy produced can be used for heating your home and a simple system will provide hot water 70 per cent of the year. Although the initial installation costs may seem a little steep, once installed, the only costs will be maintenance, and you will save up to 40 per cent on water heating, for a pay-back time of about 5–10 years.

Passive
The simplest and cheapest way of using the energy from the sun is making maximum use of the sunlight that falls on your home. This can be done by leaving curtains open during daylight and closing them promptly at dark; by having large south-facing windows and conservatories; by abandoning the use of net curtains. All of these simple measures will reduce the need to heat your home so much with central heating.

Solar Cell
The solar cells which generate are generally used in watches, calculators and in marker buoys at sea. At present the cost is very high but the long term potential is great. In the Scottish Highlands you can find solar telephone boxes, and even street lights.

Solid Fuel Stoves
These stoves are not a good choice for the environmentally-aware home. Coal pellets are produced in a way that causes pollution and environmental damage. The idea of a wood-burning stove may be appealing, but slow-burning of wood may produce carcinogenic hydrocarbons, as well as carbon dioxide and carbon monoxide. Coal fired ranges and stoves are marginally more acceptable environmentally, but the burning of coal, wood or charcoal in a domestic situation without elaborate flues and filters is not to be recommended. Keep them for decorative purposes if at all possible.

Thermostats

Fitting a proper thermostat with a time programmer could save you up to one fifth of your present fuel bill. Learn to set the timer for say, a quarter of an hour before you get up on cold mornings, so that the house is warm, and set it to shut down shortly before you go to bed at night. You can also fix times when you will be at work or away on weekends without worrying. All clocks have an override button. With a spanner and a bit of knowledge you can also fit a thermostatic radiator valve to each radiator in every room. This allows you to regulate the temperature in each room, and is especially useful when you do not use all the rooms at once.

Ventilation

A badly-insulated home may lead to great heat loss, but an over-insulated home can create problems too. Moisture has nowhere to go and can cause condensation. Bad smells and solvent fumes cannot escape and will linger around far longer. To counteract these problems we must make sure that our homes are adequately ventilated. Open the window when you are creating moisture, for example, when you are showering, cooking or washing up. Make sure that you use a kettle with automatic cut off and use tightly fitting saucepan lids. Fit an extractor fan in your kitchen and bathroom. Use natural fabrics for furnishings, carpeting, and so on. They absorb moisture and slowly release it back into the air. All of these measures will reduce ventilation problems in your home, and will provide a more pleasant environment for you to live in.

Water Tanks

Both hot and cold water tanks should be well lagged; the cold tank to stop it freezing in winter, the warm tank to keep it well insulated. Hot water tanks can be fitted with a lagging jacket that will only cost a few pounds and will take fifteen minutes to put on. Use the heat from your hot tank to warm your airing cupboard. If you are installing a new central heating system, look towards buying one in which the hot water from the taps is drawn from the mains and heated as needed. This is less expensive than heating a whole tank.

Wind Power

Windmills have been used for thousands of years for grinding corn and pumping water. The more recent wind turbines or aerogenerators work on the same principle, but produce electricity. At present there is little use of wind power in this country, but this is changing rapidly and more than fifty proposals for wind parks and other schemes have recently been formulated. The Centre for Alternative Technology has a good supply of literature on wind power, and would be able to help you with any enquiries.

Windows

These can be fitted with double glazing, but you will find that with an average cost of between £2,000 and £2,500 the pay-back period may well be the longest of all the efficiency options offered. If you are replacing the windows anyway, then it makes sense to double glaze. It may be more beneficial to use double glazing as a sound proofing against noise pollution rather than as an energy saver, but you can easily do it yourself with secondary glazing.

3 Decorating and DIY

Introduction

Doing it yourself also means taking responsibility for yourself. The craft and design in recreating and moulding a home to fit your needs increases more than just your bank balance, it also recreates your creativity and acknowledges your responsibility for the world around you.

Francis Simpson

The great British pastime – going to the DIY store - is always popular with the enthusiastic homeowner. Those of us who positively relish the idea of spending our spare time painting and decorating should also bear in mind the environmental consequences of our choices. You can champion the green cause to your benefit. Millions of homeowners are begining to realize that fixing up the home could do more than just save money or add financial value to bricks and mortar. In time, these changes could be life-savers, energy savers, pollution reducers and conservers of precious resources.

Just how well do we understand 'do-it yourself'? For most of us it generally means a quick visit to the local DIY superstore in our cars on Sunday afternoons. Many of us buy materials enthusiatically enough, but then get home and distractions help us lose interest very quickly. A smaller percentage of those who buy will endeavour to start putting up that new light fitting, clearing the landing for the wallpaper or scraping old paint off wooden window sills in preparation for that lovely new paint that had a 20 per cent discount for one week only. An even smaller percentage still will actually get round to finishing the job once started.

> **Guidelines for DIY shopping**
>
> 1. Decide on the job you need to do. Don't leave a visit to your DIY store to chance or you'll end up buying everything you don't want.
>
> 2. Write out a list of what you need and measure everything before you go in case you have to change your plans.
>
> 3. Don't forget to calculate all the extras like screws and nails, or finishing materials.
>
> 4. Phone in advance, to check that what you need is in stock, especially for unusual materials, to save a wasted journey.
>
> 5. Stick to your original list once in the store, those special offers are there so that you will make hasty purchases.
>
> 6. Don't buy materials, paints, varnishes unless they are clearly labelled.

With good management we can train ourselves to waste less, repair more and save money. No one needs to be a one-off DIY enthusiast (unless they want to be) if they start to plan a little more carefully and calculate the value and savings of their efforts to help the job have a bit more meaning.

Start with things that you can do easily. Some of these have been covered in the Energy chapter. Quick DIY jobs include fixing ordinary kitchen foil on the wall behind radiators attached to outside walls. Why not build a window-sill for those radiators that sit under windows? Any good carpentry or DIY hand book will tell you how to build a shelf and you will end up stopping all that heat from simply drifting out of the window.

Our environment is also at risk from the effects of our choice of product. Every time a house is built chemicals and finishing materials are used. The manufacture of these products has its own effect on our planet, well before they have reached the home of our choice. The quality of our world has been affected by the continued production of various materials. They can pollute rivers, killing fish and other wildlife, and they can be directly responsible for ozone depletion. In the Tees estuary in England the pollution from one manufacturing company has downgraded the quality of the estuary water so much that the salmon and sea-trout have been affected, and the German government has produced a report questioning the continued pollution of the North Sea and the Dogger Bank (an important international fishing area). Some of the toxic chemicals found have included mercury, cadmium, cyanide, phenols, ammonia, zinc, and lead.

The manufacture and export of ozone depleters for solvents and chemical materials is still happening, despite international concern. One company alone still manufactures 200,000 tonnes of ozone depleting chemicals in the United Kingdom each year. In 1988 the Department of the Environment stated that there is unequivocal evidence that chlorofluorocarbons (CFCs) and halons are causing serious stratospheric ozone depletion. Half the protective ozone layer over the Antarctic disappears every year. The United Nations has reported that even a small loss of global ozone could lead to over 100,000 more blind people and thousands of cases of skin cancer.

Timber for joints and roof spaces, flooring and beams is treated with biocides and sealed with resins. Timber treatments include lindane, pentachlorophenol (PCP) and tributyl tin oxide. These are toxic irritants, are possible carcinogens to humans, and contribute to global environmental problems. The continued production of chlorinated chemicals and pesticides for instance, adds further to the greenhouse gases in the atmosphere.

All that handy do-it-yourself work needs materials, and wood is considered a natural and renewable product. During 1988, the United Kingdom was responsible for importing 1.5 million metres of tropical sawn wood, plywood, veneer and logs. Much is used by the building industry as tropical hardwoods are a valuable and useful material. According to the Tropical Timber Federation the demand is high and still growing worldwide, especially in the producing nations themselves. Many exporting countries explain that they must sell their hardwoods for foreign exchange, often to pay back interest on loans from major banks and agencies like the World Bank and even our high street banks who have loaned money for foreign aid projects. The chapter on furniture has more detailed information about tropical forestry (see page 148).

The thousands of synthetic chemicals and materials used in the house-building industry increase the numbers of gases emitted into the home environment. Awareness of the connection between health, environment and indoor pollutants is poor because building design criteria have mostly opted for a good price and for the visual impact on the consumer rather than quality and safety. The interior designs have often been criticized because of a lack of understanding about potential pollutants from fabrics, insulation and carpets.

A common indoor pollutant is formaldehyde, found in insulating foams and chip and block boards. A floor covered with PVC lino may emit tiny quantities of vinyl chloride monomer, which is highly poisonous. Most of these commonly-found building and decorating products, especially paints, contain fungicides and lead and often various insecticides, all of which slowly release traces of toxic vapour into the air continuously. While a small amount of these chemicals will not harm you, concern has been voiced by doctors who think that over long periods, prolonged exposure may weaken the body's defense mechanisms and immunity, making other illnesses more likely.

Using cavity wall insulation to save energy poses some real problems. Urea formaldehyde foam is found in many modern insulating materials. Controversy still rages about the 'safe' levels of formaldehyde but its use has been legally restricted in several countries. One report from the United States suggested that one in five people might be sensitive to the effects of formaldehyde and similar chemical poisoning. Whether it causes long-term problems or not, formaldehyde is regarded as an irritant and any one using insulating materials, chipboards or block boards should be especially vigilant. Formaldehyde is, however, a non-methane hydrocarbon and its production contributes to the greenhouse effect. It is directly involved in tropospheric ozone formation.

While the sales of vinyl wallpapers grow, the companies that manufacture them have worked with the glue and paste manufacturers to come up with a solution to the moisture problem. These wallpapers don't allow 'breathing' or diffusion of air through the walls, you are virtually sealing yourself in. Moisture found in the air settles as condensation and this attracts mould. Hence the added ingredient in most of today's wallpaper pastes – fungicides, which of course make the product more expensive.

To avoid this problem use good quality, organic paints on walls whenever possible. This allows the diffusion required. Other, more porous wallpaper is suitable for walls that look unsightly if simply painted.

While there are many DIY problems with pollutants and toxic chemicals, doing it yourself is still the most cost effective and the least environmentally damaging way to put your house in order.

Directory

Adhesives
Solvent-based adhesives are used all over the house: to fix flooring, bathroom tiles, carpets, wood, wallpaper and plastics. Ceramic tile glues can contain formaldehyde, toluene, benzene, and similar solvents; these are often highly volatile chemicals and warning labels on tubes and packets will be strict. Benzene is carcinogenic and highly dangerous and although it does degrade reasonably quickly, it has been found in the blood of chemically-sensitive individuals in high quantities. Epoxy resin adhesives can cause severe damage to the nervous system and circulatory problems and are not recommended for people who are chemically sensitive.

Synthetic Glues
These are made from polymers like polystyrene, asphalt and polyvinyl compounds or epoxy resins and polyesters which set rigid. These are the strongest, and most dangerous adhesives found and are not renewable.

Animal Glues
These are made from the slaughterhouse wastes from animals reared to produce meat for food. They include horns, hide, bone and even fish bones. Vegans try to avoid using animal-based glues. They are water soluble and often contain preservatives.

Vegetable Glues
Examples are starches and gums; they are water soluble and are renewable.

Organic Glues and Binders
These do not contain dangerous chemicals and are available from alternative shops and green mail order companies.

Asbestos

Asbestos is a mineral fibre found in rocks. Asbestos can cause cancer of the stomach and lung; there is no safe level of exposure. Tiny fibres, too small to see, can be inhaled and become lodged in the lungs. Not all DIY products that contain asbestos are dangerous, as the real danger comes when the particles are released into the air. Asbestos cement on corrugated sheet roofing used for walls in sheds and for flat roofs can become eroded by wind and rain and literally millions of tiny fibres are released every hour. Safe alternatives exist for every use so avoid any type of asbestos.

Bricks

Bricks are blocks of fired clay. They are made from non-renewable material, and quarrying for the raw material destroys wildlife habitats. This process is very energy-intensive and the firing process produces flue gases containing toxic by-products. Bricks should be used for long-term projects only, but they are very durable and can be recycled.

Cavity wall insulation

Glass-fibre insulation has a low absorption capacity and contains small quantities of formaldehyde and polystyrene which have been implicated in long-term health problems. Natural alternatives include cork, which has been farmed, or expanded clay globules. Some architects argue that the cavity wall should be left clear and the inner leaf of the wall should be increased to compensate.

Cement and Concrete

Cement powder contains calcium, silica and aluminium. The amounts vary according to manufacture. The materials are mined from the ground, causing dust and destruction to the local area of the mining. A study of cement workers

Decorating and DIY

found that they were 75 per cent more likely to get stomach cancer. Once fixed it seems reasonably safe, although some makes do have formaldehyde and petroleum oils added.

Creosote
Creosote is made from coal tar. It is toxic to plants when newly applied and should not be used if alternative preservatives and stains can be found.

Fire
More people are overcome by smoke and inhalation of chemicals and toxic fumes from furniture, textiles and other parts of the home than are burned by fire. If your home catches fire, don't spend too long trying to put it out, you may be safer calling the fire brigade and getting everyone out of the building safely. Water fire hoses are useful to have in kitchens, as are fire blankets, especially if you are using chip pans or frying pans. Check that your fire extinguisher doesn't have CFCs – all the chemical foam types do. Take care to repair frayed electrical cords and never store petrol or similar substances in unventilated rooms.

Floor Joists
Floor joists can be left untreated, removing the risks of using pesticides and biocides, (see page 24).

Labels
Do not use products from companies that won't put a contents list on the label, even if this seems difficult. Only active ingredients are generally listed, and the intimidating names for some of them could be trade names for certain chemicals. Write to the companies and ask for a full description of the product ingredients. Also avoid products that have on the labels 'may cause cancer' or similar health warnings. In general the labels are marked with a grading scale as to their relative danger. *'Danger-Poison'* with a picture of an evil-looking skull and crossbones means that the poison could kill an adult if a tiny drop is ingested. Paraquat fits into this category, and there is no known antedote. *Warning* usually means that a teaspoon of the chemical could kill you and *Use with Caution* needs between an ounce and a litre to have the same effect.

Lead

Watch out for lead in paints and varnishes. Its use is being phased out because of ongoing pressure from environmental groups but some paints can still legally contain 0.25 per cent lead. Water pollution from leaded paint manufacturers is a major problem.

Materials

The materials you use in the fixing and maintenance of your home are of extreme importance. Can they be recycled? How long will they be useful and durable? Where do they come from?

Here is a list of the most used DIY materials in the home and how they affect the environment:

Material	Durability	Environmental Information
Aluminium	Long life	Most abundant metal available, energy intensive to produce from refined bauxite so recycling is advantageous.
Asphalt	Long life and recyclable	Solid hydrocarbon compound from crude oil waste or surface deposits. Energy to produce: minimal.
Brick	Indefinite	Burnt or fired clay, energy to produce: very intensive. Quarrying destroys local ecosystem and gases emitted.
Blockboard	Long life	Layers of solid wood from temperate sources, some tropical woods used. Bonded with formaldehyde.
Cement	Long life	Powder made from lime and clay. Extraction causes damage. Potentially toxic and corrosive to skin if additives used.

Material	Durability	Environmental Information
Chipboard	Short/medium life	Chipped woods mostly from renewable sources, small amount from rainforests. Mostly bonded with formaldehyde.
Cork	Long life	Farmed tree bark, excellent insulating material.
Fibreboards	Medium	Softwood fibres and wood residues, good smooth surface. Flame resistant treatments can be toxic.
Limestone	Indefinite	Quarried in UK, large-scale extraction discouraged.
Marble	Medium	Quarried, large-scale ecosystem damage in southern Europe. Degrades rapidly when used outside.
Metals	Indefinite	Highly energy-intensive, quarrying causing massive destruction. Some compounds are toxic and by-products and recycling could cause toxic air pollution.
Particleboard	Short/medium	Shavings, chips, sawdust and wood waste bonded by adhesives. Mostly from softwoods, pine and Douglas fir. Some use formaldehyde.
Plasterboard	Medium life	Gypsum from clays, limestone or shale. Energy intensive to produce, some is radioactive.

Material	Durability	Environmental Information
Plywood	Medium life	Wood veneers 3mm thick, cross banded to give strength. Some tropical wood used, try alder, ash, Douglas fir and birch.
Recycled fibres	Medium life	Straw, waste woods, paper etc. Used as plywood replacement or as insulating board.
Rubber	Medium	Natural extraction from trees and rubberwood is also used. Can be synthetically produced from petrochemicals but not biodegradable.
Stone	Indefinite	Any type is quarried so affects local ecosystems; best near local point of use. Large scale mining discouraged. Includes slate, granite and sandstone.

Mildew

Borax can be used to prevent mildew, added to environmentally-safe paints as an alternative to the synthetic fungicides found in many chemically manufactured brands, and to starch-based wallpaper glues.

Natural Paints

These contain natural plant oils, mainly linseed from flax and other plant substances. They also contain chalk, india rubber and aromatic oils. Organic paints can either be emulsion, water-based or gloss oil based. The best stockists are listed at the end of this book.

Paint

Paint is one of the most controversial of all the DIY products available today. In the United Kingdom we brush our way through £1.5 billion of paint, often without knowing the ingredients or how to dispose of them properly. Most paints are made from resins and fillers, disolved in toxic solvents or other materials derived from petrochemicals. Manufacture may cause extensive water pollution and some solvents used contribute to the destruction of the ozone layer. Many are corrosive to the skin or hazardous if inhaled. Use water-based, organic or 'natural' paints whenever possible, which allow the walls and surfaces to 'breathe'.

Paint Removers

Methylene chloride, a chlorinated hydrocarbon solvent, is the main ingredient of commercial paint removers especially for solvent-based paints. It is caustic and will burn skin and any surface it touches. Paint removers are known animal carcinogens and as such should never be washed into sinks or toilets. The vapours will give off phosgene gas which is extremely toxic. Use white spirit as a paint remover and use water-based organic paints, thus avoiding the need for chemical solvents.

Plaster

The inside plaster finish to a house should use gypsum plaster without foil. Phosphogypsum is radioactive, so natural gypsum is the best option. Building biologists recommend sand and lime as another option, provided it is finished with a smooth overcoat of wet gypsum plaster. Cement-based plasters are often used, but they are very dense and therefore don't encourage diffusion or moisture absorption.

Safety Guidelines for using Dangerous Chemicals

If you feel that you simply must use toxic chemicals in your home, like pesticides, preservatives for wood or solvents for any reason, the least you can do is follow a set of guidelines that protect you as much as possible:

- Always check labels on bottles and cans before you buy, make sure that you know exactly what you are buying;
- Never use any dangerous chemical product for anything other than the use for which it is sold;
- Never smoke while applying these chemicals;

- Keep them well away from all food sources, if you have to use them in your kitchen for instance, take all food out of the room before you start to use them;
- Use strong, industrial-type gloves and protective clothing;
- Avoid splashing, spilling and breaking containers, make sure you have the right equipment with you to help you;
- Wash your hands and clothing after applying;
- Never mix chemicals together unless the instructions tell you to do so;
- Store dangerous chemicals safely, never transfer them into unmarked containers, and keep them completely clear of children;
- Dispose of empty or half-filled containers with extreme care. Poisons should never be tipped into gutters, sinks or drains. Call your local environmental health officer at the council offices who will give you instructions on various chemicals that need to be disposed. They can sometimes collect them for disposal.

Septic tanks

Many people in country areas don't have access to the main plumbing facilities and have septic tanks to collect any waste water. Often, householders are more careful about what they put down the toilets and in waste water because toxic or foreign material may kill off the natural bacteria that break down the sewage. Sanitary towels, disinfectants, bleaches, paints or any toxic or potentially poisonous materials are definitely not recommened. A natural septic tank activator is a quarter of a pound of active yeast with a pound of brown sugar mixed with four cups of warm water and flushed down the toilet.

Shellac

This is the resinous secretion of the lac insect. It is one of the natural varnishes available. It has excellent sealing qualities for wood. It is not suitable for floors or surfaces which are subject to heavy wear and tear however.

Solvents

Solvents or volatile organic compounds are used in paints, cleaning solvents, polishes, plastics and liquid heating fuels.

They readily evaporate in the air and cause air pollution. They are the smells that fill the room while decorating and painting with petroleum based products. Most are asphyxiants and intoxicants. They can cause dizziness, disorientation or as the result of a large dose, death. Choose paint products which are water-based and which contain organic ingredients.

Timber Treatments and Preservatives
The London Hazards Centre argues that the need for chemical timber treatment comes from bad design practice even when the timber is exposed. Other research has shown that chemical timber treatments are no more effective than leaving the timbers bare and that the guarantees given by these companies are speculative. Timber should be carefully chosen for the job, correctly conditioned before installation and used properly in the building. When positioned, the timber should be well ventilated. Chemicals used in timber treatment do not remain inert within the wood, but become gases, and these are released into the surrounding environment, sometimes over long periods. The toxic side-effects that this could cause are numerous. If you need to use a timber treatment do some thorough research first; these are some of the most toxic chemicals that you can bring into the home.

Tool Kit
A basic tool kit should include:

- Hammer
- Screwdrivers
- Assorted nails and screws
- Adjustable wrench
- Saw (various sizes)
- Measure or rule
- Spanner
- Chisel
- Sandpaper
- Pliers
- Adjustable blade knife
- Drill (hand or electric)
- Spirit level
- Mallet
- Rawlplugs
- Tacks
- Fuses
- Insulating tape
- Paint brushes
- Spare plugs

This basic tool box should allow you to do most of the do-it-yourself jobs around the house. Think of all that money you could save and then think about saving resources for the planet too!

Wallpaper
Although painted walls are easier to keep clean than traditional wallpapers, many people are opting for vinyl or plastic varieties which can be washed but which stop your walls breathing and need wallpaper glues containing fungicides. Millions of trees are cut each year to satisfy our demand for wallpapers so if you need to use them choose a design that will last and is as durable as possible.

Whitewash paint
One of the least complex types of paint available, made from lime mixed with water, it is safe to use for people with sensitivities but is not particularly durable.

Wood Treatments
Wood only needs to be treated inside with linseed oil or beeswax, and exterior wood can be treated with stains or oils and then varnished with a natural or linseed oil base. This means the wood is able to breathe, and you can feel its wonderful texture and surface. Waxed floorboards render them a semiconductor, which is the ideal surface for touch by humans.

4 Flooring and Fabrics

Introduction

To create a tactile and textured room, the choice of fabrics and carpeting must be fresh and lively, natural and warm. The more natural the final choice, the warmer and friendlier the final appearance.

Elizabeth Whitefield

Fabrics used for clothes and as floorcoverings have been around for at least 20,000 years. Pictures of old clothes can be seen in the cave paintings of France. Comparatively speaking, weaving and knitting are fairly recent inventions, having been introduced only 7,000 or so years ago. Silk worms have been farmed since about 2,000 BC.

All fabrics have an environmental impact, often because of how they are produced, not just because of what they are produced from. If you look around your home today you will see a wealth of styles, textures, shapes and weaves. They keep us warm, add colour and creativity to our lives and provide tactile surroundings. Can you imagine your home without them?

Around the house, fabrics and furnishings come in all shapes and sizes. Carpets and flooring are important for our feet, and for their aesthetic value. Originally carpets were draped over furniture or hung on walls. They were used as soundproofing and provided basic insulation. With the introduction of central heating, their function has become more decorative than functional for many homes, although the basic warmth from carpets will never be lost. Carpet making as we know it originated with the nomadic tribes of the Middle East regions. Where sheep and goats are kept for food and wool. The region provides many of the world's best known dyes, resulting in the beautiful shades now associated with Persian rugs. More recently these dyes have been replaced with synthetic chemical dyes, derived from petrochemical sources.

Wool has been produced in this country for centuries. In the Middle Ages the importation of cotton was forbidden because of arguments over the vulnerability of the wool farming industry and the weavers. These days wool has become just another of the dozens of fabrics that are imported and used for clothing and textiles in our homes. Australia now produces one third of the world's wool and satisfies over 70 per cent of the United Kingdom market.

To produce wool in the vast quantities now used by industry, sheep rearing is done on a massive scale. By law, sheep are required to be dipped, usually in an organophosphorous liquid, to rid them of fleas and mites. Unfortunately however, this produces an environmental hazard from the toxic by-products from the wool-processing industry. Fly-strike is a continual problem; blowflies lay eggs in the sheep's wool and the larvae attack the flesh of the host animal. The pain that the sheep can experience is immense and the sheep can die within a week from stress or blood poisoning. Mortality rates in Australia are particularly high and during one of the worst years, over 3 million sheep died of fly-strike. In Australia the flies have become resistant to all but the newest chemical treatments. One particularly barbaric treatment which is carried out on about 80 per cent of the sheep reduces the risk of fly-strike but removes layers of skin from the sheep round the breech and tail area, without anaesthetic using sharp farming shears. The mutilation is known as mulesing and is not done by veterinary surgeons but by the farmers themselves.

When sheep are chemically treated the fleece can contain significant concentrations of the organochlorine residues from dieldrin and hexachlorocyclohexane (HCH) which are washed off in the manufacture by the wool processing industry. The British Textile Technology Group have studied the effluent – or waste water – from many wool processing plants, and found that raw wool scouring produced significant discharges of the chemicals into rivers and water systems. Dieldrin has come under particular scrutiny recently and has been officially banned in the United Kingdom but

Flooring and Fabrics

gets back into our environment through imported fleeces.

Although these chemical treatments are thought to be carcinogenic, the sheep don't seem to be immediately affected from the dipping. They are often killed within a short time for food in any case. A significant proportion of the wool is skin wool obtained from slaughtered sheep. 25–50 per cent of slaughterhouse profits come from sale of hides.

The whole textile industry in the United Kingdom is under increasing pressure from water and pollution authorities over the pesticides that it discharges into effluent. The industry claims it has little control over imported stock but legislation from the European Community has set strict guidelines for pesticide discharge. One of the major polluters results from the use of pesticide pentachlorophenol (PCP). It is used as a rot-proofing agent to treat cloth before being exported. Some of the pesticide is washed off during the finishing process and is discharged into the sewers where it finds its way into rivers. The rest of the pesticide stays in the cloth where it can be absorbed through the skin over time.

Carpets can be a breeding ground for microscopic insects and fungus; wall to wall carpeting and central heating is a combination which encourages this. Our vacuum cleaners are not sealed enough to hoover up these pests – 40 per cent of the dust they suck in goes out through the ventilation hole, so the bugs continue to multiply in the carpets. For people with allergic reactions this is a real problem. Flooring such as linoleum offers a solution because it can be mopped clean. But this is hard and sometimes cold, so the addition of cotton rugs can be both decorative and warming.

Textiles like linoleum have begun to be replaced with synthetically produced products. Since the end of World War Two hundreds of synthetic fibres have been developed, and over 25,000 barrels of oil are used each day in the production of synthetic materials. There is little scientific evidence which proves conclusively that they are harmful. However, chemicals such as phenol which is toxic, and vinyl chloride which is carcinogenic, are used. Even if not present in the fabric the workers who make these fabrics will be exposed to the chemicals. They can also be derived from hydrocarbons, glass and graphite. Polyethylene terephthalate, also called PET, is another plastic which is used for soft drinks bottles.

There is a possibility that these could be recycled into terylene trousers!

These plastic fabrics are not very comfortable as clothing. They do not absorb moisture well, providing an ideal environment for bacteria to grow nor are they good conductors of heat. They are difficult to clean as they absorb oils from the skin and retain oily stains. These can only be removed with special synthetic detergents using enzymes and surfactants. Static is another common problem when using synthetic fibres. The solution to this problem is to use even more chemicals in fabric softeners and antistatic agents.

Natural fabrics come from abundant and renewable resources and are usually produced as annual crops. They are comfortable to wear and they are also biodegradable – but this does not mean that they are environmentally benign.

Cotton is part of the mallow family. It is a shrubby plant that only grows in hot climates. To grow the plant takes vast quantities of water. Water is very often scarce in these countries, so precious supplies are used for cash crops like cotton rather than food. As a cash crop cotton occupies 5 per cent of the world's productive land. It is an extremely difficult crop to grow because of root rot and pests such as pink vole worm, cotton mouse and white fly. A very dangerous cotton pest is the boll weevil which can take only five days to reproduce and can quickly destroy an entire crop. Chemical pesticides, instead of natural systems, are used to control the pests, and these in turn pollute the environment.

In Egypt they use nearly 30,000 tons of pesticides on cotton alone. These pesticides contaminate the water supplies and cause numerous health problems for the people living and working nearby. In Sudan, pesticides used on the cotton crop contaminate drinking water, which is becoming too polluted to drink. Fertilizers are used on the cotton in Brazil which turns the cotton pink and even more synthetic chemicals must be used to make it white again. We need to ask for unbleached cotton which has been organically grown. In the United States there are a few suppliers, but it is impossible to find in Britain – with consumer power it could become plentiful.

Researchers in the Californian Department of Health Services surveyed 406 residents in six small agricultural communities. People living or working near the sprayed fields were found to have 60 to 100 per cent higher incidence of various self-reported symptoms such as fatigue, eye irritation, nausea and diarrhoea. Roughly one hundred pesticides were reported to have been used on the cotton in the study area during the defoliation season.

While the texture and colours of natural fabrics are unlikely to be matched by any synthetic product, we should be conscious of the impact on the environment that all these materials have and use them wisely.

Here are some simple guidelines for buying and using fabrics

1. Read the labels on all fabrics. Buy only 100 per cent natural fabrics.
2. Make sure that backing for carpets are natural.
3. Wash all fabrics before use in soft natural soap.
4. Make better use of bare natural wood, such as for floorboards, and alternative fabrics like hessian for halls and stairs.
5. Use curtains with linings to help with heat loss and insulation.
6. Re-dye fabrics to brighten them and increase their lifespan.
7. Take unused or secondhand fabrics to charity shops so they can be sold or recycled.
8. Use home-made starch, fabric softeners and moth repellents.
9. Make your own clothes, curtains and upholstery coverings.

Directory

Bleaches
Most fabrics have been bleached before they are sold. Buy fabrics direct from the mills and bleach naturally in the sun-

light. They can then be dyed with natural dyes. For large pieces an old-fashioned tub washing machine is very good.

Carpets

Carpets should be made of 100 per cent wool and backed onto jute or a similar fabric. Synthetic carpets cause static, which upsets the body's electrostatic energies and are made from chemicals which will give off gases for most of their lives. Some people are allergic to the chemicals used to rid carpets of moths. Foam backed carpets contain CFCs.

Where possible, carpets should be made of a long-lasting and natural source like wool, should be backed by hessian and the underlay should be natural, like felt. Wall to wall carpeting can be the breeding place for pests and other microscopic organisms.

Coir

This is the fibre that comes from the outside of the coconut. It is strong and hardwearing, used for matting and rope. Coir made from coconut matting is very good for halls and rooms or in heavily-used areas. Coir can now be bought with a latex backing which increases its life considerably. Sisal and seagrass are similar products, seagrass being very useful in bathrooms.

Cork

Cork is made from the outer bark of the cork oak tree. Cork bark regrows itself after stripping and all of the bark is used without waste. It is made into floor and wall coverings and tiles. It requires little or no treatment except when used on floors. Check that it has not been backed with a synthetic covering.

Cotton

The cultivation of cotton requires vast amounts of water, pesticides and fertilizers. There are few sources of organic cotton but many companies are producing raw cotton which is a natural shade and has no bleaching agents used to make it white. Cotton has the advantages of being renewable, recyclable and quickly biodegradable. As its growth and production uses a lot of energy, it should be recycled whenever possible. Manufacturers who use synthetic threads to sew cotton together hamper the recycling process, as the different fibres cannot be separated easily.

Dry Cleaning
Dry cleaning is the use of detergent and chemical solvent rather than a detergent and water. One dry cleaning fluid is perchloroethylene. The inhaling of this gas can cause liver damage and cancer at worst, or light-headedness and nausea from small amounts of exposure. Try to buy clothes that do not require dry cleaning, i.e., natural fabrics. Most fabrics can be washed. For delicate fabrics, handwashing is best anyway. Linen can be treated like cotton but should be ironed when damp. Wool should be washed in soap and reshaped over a towel and left to dry flat. Don't dry clean your duvet as the chemicals will impregnate the interior fibres and you will breathe these in as you sleep. They can be washed in the bath and left to dry in the sun.

Dyes
Chemical dyes used for cotton and other natural fabrics can be irritants for the sensitive and toxic for the employees who work with the dyes. Reactive dyes offer the textile industry a wide range of bright colours which are resistant to fading. The handling of the dye powders and solutions before application can give rise to dust or fine airborne pollution which can be hazardous if inhaled or when in contact with the skin.

The Health and Safety Commission states that there is no known hazard to anyone handling or wearing textiles which have been dyed with reactive dyes at present. Direct dyes contain benzidines and dichlorobenzidine and concern has been raised in the United States that these carcinogenic compounds can be absorbed into the skin. They suggest avoiding wearing clothes dyed with deep colours, or washing them before you wear them.

Natural plant dyes are an alternative.

Colour	Alternative colour source
Dark brown	Bracken, sorrel, walnut shell
Tan	Elder, birch, broom, heather, bedstraw
Blue	Indigo, woad, elder
Black	Gall nut, oak bark
Green	Gorse, iris, broom

Colour	Alternative colour source
Purple	Blackberry
Yellow	Apple, hazel, bracken, gorse, bedstraw
Orange	Ragweed, bramble
Magenta	Dandelion
Red	Bedstraw, madder, tormentil, foxglove, cochineal, lichen

Natural dyes require the use of equal plant material to fabric or fibres to be dyed so make sure that you are not using any rare plants for the dyeing process. Lichens are a particularly good dye source, but some lichens take up to one hundred years to grow, and they are being destroyed anyway by airborne pollution. Different mordants will give different shades and colours. The basic mordants used in dyeing are alum, tin, chrome and iron. Many mordants are poisonous, so handle them with care, especially when throwing the dye bath away. There are two basic methods of dyeing yarn: one uses abjective dyes which only need to be dyed with the yarn to produce a fast colour such as woad. Subjective dyes are only colour-fast if the yarn is first impregnated with a mordant. With this method the yard is wash-boiled with the mordant and then immersed in the dye bath. Cochineal and lichens need a mordant.

Fabric Finishes
Many cottons have easy-care and non-crease finishes. This means that they have a permanent formaldehyde finish which does not come out with washing. The fabric is easier to keep: it doesn't need to be ironed for instance. Fabric finishes usually contain formaldehyde. Polyester cotton sheets have a particularly heavy finish because of their frequent washing. Nylon fabrics could have a formaldehyde finish to make them flame-proof. The labels will reveal these finishes saying crease resistant, permanent pressed, non iron, shrink proof, permanently pleated, waterproof. The finishes combine the formaldehyde resin directly with the fibre, making the formaldehyde irremovable. The people most at risk are the clothes packers and pressers. Formaldehyde is a contributor to the greenhouse effect through chemical reactions in the atmosphere, and is an irritant in small doses and an allergen to many people.

Feathers and Down
Gathered from the natural moulting of birds, they are the traditional filling for duvets and cushions. They are washed, as air drying chemicals would destroy their natural insulating properties. Eiderduck down is the lightest and the most expensive. Feather fillings vary in quality and many people are allergic to them. Duck and geese need not be killed to produce their feathers, but most are.

Flame Resistant Fabrics
Formaldehyde is one of the chemicals used for this process. This can leave formaldehyde on the surface of the fabric, which could cause skin irritations. In March 1993 legislation will require all furniture including bedding and mattresses to have a fire resistant covering or a fire retardant interlining. Natural fibres burn at a lower temperature (300 degrees F) than synthetics but they do not give off potentially lethal gases. Wool is naturally flame resistant. In the United States wool mattresses and futons are exempt from the fire regulations. If you do not want to sleep on one of these mattresses you can put a fleece underblanket on the bed.

Hemp
The coarse fibres of this plant are extracted and used for matting rope and cloth. However, the use of this annual plant has declined due to the plant being the source of cannabis and its cultivation is now strictly controlled.

Jute
The fibres are extracted and mixed with hemp to make hessian, which is used in sacking and wall coverings and also as a natural backing for linoleum.

Kapok
The seed pods of the tropical tree produce silky fibres with a soft texture. Not suitable for spinning but because of its thermal and waterproof qualities used in mattresses and cold weather clothing.

Leather
Leather tanning processes sometimes contain formaldehyde and chlorinated chemicals. There are natural tanning pro-

cesses which can be used. They use plant extracts like tea and natural oils, and the by-products can be used as garden fertilizers. Leather is also naturally quite flame resistant. Leather is part of the meat industry, a multi-million pound business which is boycotted by vegans.

Linen
The production of linen uses a process of natural fungi and bacteria, the non-creasing finish is petroleum-based. Linen is the best environmental fabric. Linen is made from fibres from the inner bark of the flax plant. The bark is first removed from the plants and left in the field so that natural bacterial action can loosen the fibres. The stems are then crushed mechanically and the fibres removed.

Linoleum
'Lino' is a very good natural floor covering which can be used throughout the house. It is easy to clean so therefore more hygienic than other flooring. Traditional lino is made from powdered cork, linseed oil, wood resin and wood flour mixed with chalk and pressed onto a backing of traditional hessian or jute canvas.

Moth Balls
Some chemically-produced moth balls are made from a volatile chemical called paradichlorobenzene. The gases released from the balls saturate clothing over long periods, causing irritations to the nose, throat and lungs when breathed in over a long period of time. Many herbs are natural repellents to insects and animals and can be used instead of moth balls. These include lavender, rosemary, mint, peppercorns and cedar wood. For storage of woollens over the summer, put in air tight boxes containing these herbs. The herbs can be mixed together in cotton or muslin bags.

Parquet Flooring
Large amounts of wood block flooring come from tropical forest sources and should be avoided. Some reputable sources are found in Europe which use only sustainably-produced timbers. These natural floorings have the advantage that they do not contain any harmful chemicals in their pro-

duction, some companies even claiming to use non-solvent sealers and adhesives.

Pesticides
Most fabrics will contain a residue of pesticide. Organophos-phate and organochlorines can be absorbed through the skin. Soak all textiles in hot water with baking soda, vinegar or borax before use, and wash thoroughly. Especially bed linen.

Rattan and Bamboo
Used in furniture, usually grown wild in the tropical forests. Problems start when it is prepared for export. It can be sprayed with sulphur or rangoon oil. The oil gives bamboo the dark flecks. Most often treated with a mixture of DDT and coconut oil. For the healthy home treat bamboo with an organic varnish, or shellac, both found in good DIY stores.

Rayon
Rayon is made from the cellulose fibres in eucalyptus and other plant-based fibres such as seaweed, peanuts, and maize. It mainly comes from the eucalyptus tree. It requires chlorine in the early stages of production. This in turn causes dioxin and other organochlorine pollution. These are toxic compounds found in our water supplies which pollute the food chain. Exposure to organochlorines increases the pollution load on our bodies.

Also, a eucalyptus tree can soak up to 400 litres of water every day. Women in northern Sumatra, Indonesia, who argued that their precious water resources and land had been destroyed by eucalyptus plantations were jailed for destroying 16,000 eucalyptus trees which had been newly planted for rayon production.

Reeds
Each country has its own variety. Most reed used in this country is grown in Holland. Can be used for roofing-thatching, mats, baskets, chair seats and can also be used to make brushes and brooms. Seagrass is imported from China and Taiwan and is used for flooring. After cutting it is dried and does not contain pesticides.

Rubber and Latex

Rubber is a very strong, waterproof and popular material, derived from the sap of the rubber tree. It can be bought in sheets and tiles. It is a sustainable product grown mainly in Malaysia and Brazil.

Rush Matting

This type of flooring is very good for damp areas of the house, since they like to be slightly damp. In warm weather they need to be watered with a watering can. It is loose laid and so not so durable; probably best used in the bathroom and toilet.

Seagrass

This is made into a soft textured but durable matting for the home. Its production does not involve pesticides and fertilizer but its natural habitat – the wetlands – is under threat from reclamation for agricultural uses. Pesticides also leach into the water in these areas from the surrounding farmlands.

Silk

The production of silk requires the asphyxiation of the silk worm. Its production is land intensive, using 1 kilo of silk for every 200 kilos of mulberry leaves. Look for alternatives, like rayon, made from mulberry cellulose.

Stains

There are many ways to remove stains on clothing and fabrics without using synthetic chemicals. Here is a quick guide to the most common spills and stains and how to remove them:

- Tea: remove with soap and cold water as soon as it happens.
- Red wine: put salt on the stain.
- Candle wax: iron with brown paper.
- Blood: put in cold water immediately.
- Perspiration: soak in a solution of water containing white vinegar or bicarbonate of soda.
- Rust: lemon juice and salt rubbed on rust stains which are then left in the sun is very effective. Sun light both

whitens and deodorizes. In hot summers this is worth trying on all stubborn stains.
- Coffee: rub stain gently with a little glycerine and wash with tepid water, then iron with moderate heat on the wrong side.

Synthetic Fabrics
Polyester, viscose and acrylic are some of the most common names that you see on labels. All are non-organic and are non-biodegradable. There is some evidence that the chemicals that are used in the production of these fabrics irritate the skin and the environmental effects of using petrol-derived materials are immense. Synthetic fabrics are said constantly to give off fine fibres which float in the air. Airborne pollutants of this kind are said to contribute to the rise in child asthma and allergies.

Synthetic Flooring
Synthetic flooring is produced from petrochemical derivatives, polluting in their manufacture and of course at source when the oil is extracted. They are not biodegradable and have never been recycled, although at least one company in the United Kingdom claims to use recycled plastics in their production.

Vegetable Fabrics
Vegetable Fabrics; these are made from grasses and various tree barks and the husks of coconuts. Investigation is being made into this area for renewable sources.

Wooden Floors
Wooden floors are very beautiful if cleaned down and waxed with a natural polish. In old houses waxed floorboards can be a lovely alternative to carpets. If the boards need to be sanded, make sure there is adequate ventilation while working, and wear a mask. Properly finished timber flooring can last for hundreds of years and is very recyclable. It can be varnished or waxed with natural products. (Although one problem is that many mortgage companies require that timbers are treated with toxic pesticides and fungicides.) They look impressive when varnished and are easy to keep clean. For warmth, cotton rugs can be placed where you sit.

Wool
Like all natural fibres it can be reused and recycled, as it has a very long natural life. Problems arise with the treatment of sheep and chemical pesticides used in dipping (see page 60). Unfinished wool has very good waterproofing qualities. Wash before use to remove as many chemicals as possible.

Wool Sources
Other wool sources include Kashmir goats, angora rabbits, yaks, llamas and alpacas. The coarser wool of the angora goat is mohair, and often mixed with lambswool. It appears that the wool comes from the shedding of the animals in the spring or shearing, but the hair is a by-product of meat and leather production. It is a matter of personal discretion as to whether these products are bought. Eventually these animals will be killed for meat, and very often do not have a natural life while they are producing the wool for human use.

5 Cleaning

Introduction

All too often keeping somewhere clean simply shifts pollution from one place to another.

Herbert Giradet

Our kitchens, bathrooms and living rooms are not generally associated with environmental pollution, except when we think of the dirt that we generate around the house! But it is becoming increasingly clear that some cleaning products themselves are creating problems for the environment in their production and use. Under our sinks and in our bathroom cupboards sit a myriad of cleaning products which we have bought to cater for our every need. It seems that we need to buy more and more specialist products too. Have you ever wondered why we now appear to need a dozen products to do various cleaning jobs when we once managed with just one or two?

Pollutants associated with Cleaning
Environmental problems range from water pollution by phosphates to the testing of various products on animals and the unnecessary cruelty that this causes. The weekly wash by over 90 per cent of households use washing powder that has been tested on animals and includes phosphates, enzymes, surfactants, bleaches, and chemicals like EDTAs. EDTA, or ethylene diamine-tetra-acetic acid is an additive needed in soap powders to stabilize bleaches, like the sodium perborate used by most companies. This additive combines with toxic heavy metals in the water like cadmium, lead and mercury, and may help them get back into our

drinking water supply. The bleaches themselves are often irrelevant, as only one in four of our washes ever needs any kind of bleach. As about 20 per cent of the powder content is bleach, this means that it is wasted in vast quantities as well as being an environmental hazard. Optical brighteners are also added to ordinary soap powders; but they only give the impression of cleanliness – by adding tiny particles which glow in ultra-violet light and make clothes *look* brighter.

Phosphates constitute up to 30 per cent of the content of your average washing powder. This is one ingredient that has come under severe criticism from environmentalists in recent years. In the United States, federal regulations ensure that the maximum allowable percentage of phosphates in a box of washing powder is never more than 4 per cent because of the proven harm the phosphates cause. They are accumulating in lakes and rivers, and the result is a build-up of algal blooms which suffocate all other life forms in the water. When the water from your washing machine is mixed with other industrial waste and sewage, a chemical cocktail gets washed away into our seas at the rate of millions of tonnes each day.

Other cleaning products have caused environmental problems. Perfumes, colourings and other synthetics like freshening agents are added to a variety of products, including hair shampoos, carpet cleaners, deodorants and antiperspirants, fabric shampoos and cleaners, and hairsprays. A large number of these products contain various hydrocarbons, known as percursor greenhouse gases. These gases contribute to the greenhouse effect and are directly involved in urban smog and tropospheric (lower atmosphere) ozone formation. These chemicals include formaldehyde, acetylene, ethylene, ethane, and butane.

Some air fresheners and fabric sprays work because they block the sense of smell, attacking the tiny hairs in the nasal passage. This means that you cannot smell anything else except the lemon or woodland smell that is presented. This type of cleaner could have disastrous effects; you can no longer smell bad food, and you may not notice gas leaking, for instance. The best way of getting fresher clothes or rooms is simply to open windows, keep the rooms as clean as possible and watch out for rotting food or similar problem areas.

Cleaning and Human Health

Some cleaning products are based on much more corrosive ingredients, like sodium hydroxide in oven cleaners, or sulphamic acid sometimes found in kettle descalers. These are toxic if swallowed, and often have a very dangerous vapour if breathed. Descalers to rid limescale from your bath or toilet pan can also be toxic. Those with acidic properties should never, ever be mixed with bleaches like sodium hypochlorite – the most commonly-used liquid bleach – as the mixture frees the chlorine into chlorine gas, which can be fatal if inhaled. Often these products can severely damage the skin or eyes if you come into contact with them; imagine what they do to wildlife when you pour them down the sink or toilet!

Trichloroethene is one of the commonest solvents used in products such as stain removers. It is sometimes known as methyl chloroform, and it is known to deplete the ozone layer. Trichloroethene is also very persistent in water, and has been known to cause damage even in very small concentrations. Solvents are also dangerous if they are accidentally or deliberately inhaled. Large doses can lead to heart failure and long-term effects on the central nervous system.

Over 100,000 chemicals are used worldwide by industry every day. The majority of them have not been adequately tested for risk to our environment or even for risks to human health. Some cleaning products contain chemicals which may be carcinogenic to animals and humans, some may be releasing tiny quantities of toxic chemicals into our bodies and environment over long periods and may increase our chances of falling ill with chemical sensitivities or allergies. Synthetic chemicals can cause cancer, miscarriages and lead to birth defects when they are misused or used in excess, and the sophisticated technology which can monitor some of the by-products and long-term hazards of the industry is only just becoming available.

There is much confusion about which chemicals used in the home are dangerous and which are not. It is certainly impossible to avoid chemicals completely, but you can reduce your exposure to synthetic cleaners by opting for simple, non-toxic alternatives. Most household cleaning requires nothing more than a simple solution and a bit of physical effort! Our obsession with cleaning has been further encouraged by advertising and marketing. Women are especially targeted by industry as they are often responsible for buying the cleaning products and doing the actual work. Adverts imply that women are not good wives or mothers unless they buy certain products. Children are often used to emphasize a mother's uncleanliness or the need for super-clean clothes, floors or toys and yet an average toddler eats about a tablespoon of dust and dirt each day without suffering too many upsets! Advertisements use powerful language to entice us to use the strongest solution possible, killing germs dead and getting into all those inaccessible corners. The emphasis is often on the danger of dirt rather than the dangers of the chemicals used to clean them.

The market for cleaning products is owned by a few companies, although you will see dozens of brand names on the shelves. Most soap powders for instance, are manufactured by two major corporations: Proctor and Gamble and Lever Brothers. They operate internationally and produce similar products on both sides of the Atlantic; sometimes they change the name but keep the same ingredients, sometimes they have to adhere to local or national laws and regulations affecting the labelling or the content of the product itself.

Hints for Buying Greener Cleaners

1. Stick to simple, non-toxic products, like bicarbonate of soda and vinegar, as often as possible.

2. Avoid overpackaging – ask yourself can this package be recycled or refilled?

3. Avoid pesticides or herbicides to kill flies and other insects.

4. Use the smallest possible amount of cleaner to get the job done.

5. Resist the advertising – do not buy specialist cleaners.

6. Buy cleaners and toiletries that have not been tested on animals.

7. Write to those manufacturers who refuse to list all their ingredients.

8. Avoid aerosols.

9. Try a dustpan and brush sometimes instead of using the vacuum cleaner every time.

10. Keep any dangerous chemicals securely stored away from children and dispose of them very carefully.

11. Only put biodegradable products into your toilets and sinks.

12. Do your cleaning in bulk – clothes must fill the tub, and dishes should be washed together.

Directory

Aerosols
Do not buy any cleaning materials in aerosol containers, even if they claim to be 'environmentally friendly' (over 1 million are still bought each day). The chemicals replacing CFCs may not affect the ozone layer, but they do contribute to the greenhouse effect. They are also highly volatile, and the pressurized hydrocarbons must not be placed near a heat source or in direct sunlight, as they are likely to explode.

Air Fresheners
Commercial air fresheners are potentially dangerous. Many irritate the mucal lining of the nasal passages, and some even block the sense of smell by deadening the olfactory nerve! Most are produced in aerosol cans, which are a poor form of packaging and an environmental hazard (see Aerosols). Opening your windows, and banning smoking are simple ways of reducing household smells. Use bowls of pot-pourri, scented candles and essential oils as natural air fresheners.

Airing
Airing clothes can help the environment by cutting down on the need to use dry cleaners (see page 80). After wearing,

brush clothes gently and hang them out to air (out of doors if you can trust the weather). Sponge shiny areas with a weak salt and vinegar solution, and you will be amazed how fresh your clothes will be.

Bathing

Instead of using expensive (often allergenic) commercial bath products, use a few drops of your preferred perfume and a couple of handfuls of bicarbonate of soda, sea salt, or Epsom salts. Or imagine yourself as Cleopatra and relax in a milk bath. Just put a couple of teaspoons of oatmeal and powdered milk in a muslin bag, a few drops of essential oils, and soak away.

Bicarbonate of Soda

Probably the best product you can have in the house as a cleaning agent, because it serves so many purposes. It is a scouring powder, a polish and a cleaning agent. It can even act as toothpowder and fungicide. And just think how much more space you will have in your cupboards when you dispose of all the unnecessary products that 'bicarb' will replace.

Borax

Borax is another must for cleaning in the green home. Forget those commercial stain removal products, a solution of one part borax to eight of water will deal with a multitude of stains, including, blood, coffee, mildew, mud and urine, at a fraction of the cost. A hospital in the United States used borax instead of commercial disinfectant for a year. They monitored it carefully and found that it met all necessary germicidal requirements, and saved them enormous amounts of money. Borax is available from your chemist, but should be treated with caution as it is toxic.

Bran

Delicate fabrics such as lace can be washed in bran water, which is a very gentle cleaner. To make it, fill a muslin bag with a teacup of bran. Boil the bag in six pints of water for about ten minutes. Squeeze the bag when the water has cooled, and you have fresh bran water for soaking and washing.

Brass
Use lemon juice or white vinegar mixed with bicarbonate of soda to clean brass. A wonderful tip (from Mrs Beeton), is to save haricot bean water and immerse brass hooks, handles, and so on in it to clean them up beautifully.

Brooms
Dip a new broom in hot salt water solution to toughen the bristles and give it a much longer life.

Brushing
Often a good old-fashioned brushing of clothes will remove superficial stains. Regular brushing will also keep fabrics in good condition.

Carpet Cleaning
Try shaking baking soda or cornflour on carpets, or 1 part borax to 2 parts maize meal (which is all expensive carpet cleaners generally contain), leave for an hour or two and then vacuum. For more stubborn stains use washing up liquid or white vinegar mixed with boiling water.

Containers
Try to be practical about using containers for your home-made cleaning agents. Dry powders such as bicarbonate of soda could be placed in an old talc container, jam jar or flour shaker (carefully labelled). Solutions could be stored in plant sprayers to use on stained carpets, for cleaning windows and so on.

Copper
To keep copper kettles shiny, fill them with hot water and polish the outside with a rag dipped in buttermilk or sour milk. Copper also responds well to a rub-over with lime or lemon juice and salt. Polish it up with a duster.

Cutlery
If you have silver or silver-coated cutlery keep a small piece of chalk in a drawer or canteen. The chalk absorbs moisture and prevents the cutlery becoming tarnished so quickly.

Dandruff Shampoos
These may be made from poisons including selenium,

formaldehyde and cresol, all of which are absorbed by the skin and can cause tiredness, headaches, and can even burn eyelids. They also cause pollution problems when they are released into our water system after use. Try washing your hair each day with a tiny amount of shampoo. If that fails to work you could try massaging handfuls of bicarbonate of soda into your hair after washing it. Remember to rinse it very thoroughly, or you may be left with very stiff hair! You could try improving your diet (dandruff may be diet related) or lowering your stress levels (dandruff may be a symptom of stress) or you could just relax and you may discover that it goes away naturally.

Dishwashing Powder
If you own a dishwasher, you can make your own dishwashing powder by mixing 1 part borax with 1 part bicarbonate of soda. That way you are guaranteed that it does not contain strong chemicals and was not treated on animals.

Doormats
Good doormats will save you time and energy spent cleaning your carpets. It is said to be the abrasiveness of dirt which causes wear and tear on carpets. To clean and condition coconut fibre doormats, beat them face down outdoors, then clean them with hot water and washing soda applied with a stiff brush. Finally, give it a salt water rinse to keep the fibre stiff.

Drains
Washing soda crystals are a good choice for cleaning drains. They are much kinder to the environment than toxic bleaches and disinfectants. Equip yourself with a good plunger as well and you should rarely have to call out a plumber.

Dry Cleaning
The dry cleaning industry has a poor record when it comes to taking care of the environment. Many of the chemicals used have been proved to be carcinogenic. The two most commonly used solvents, trichlorotrifluoroethane and perchloroethylene, are organochlorines known to pose a severe health threat. The first also contains CFCs, and therefore contributes to ozone depletion. The best advice is not to buy

clothes that need dry-cleaning. If you do have 'dry-clean' only garments, try washing them by hand with very mild soapflakes. The majority of clothes can be cleaned this way without any ill effects.

Dustbins
If you burn a couple of newspapers in a metal dustbin, it will absorb grease, dirt and unpleasant smells. Placing a couple of handfuls of straw (try your local pet shop) in it will have the same effect. If you have a plastic or rubber dustbin, wash it out with a washing soda solution, and sprinkle bicarbonate of soda in the bottom.

Eucalyptus
Eucalyptus oil can be used to remove grease stains from the most delicate of fabrics without leaving a trace. It also works on oil and grass stains. Rub it in well and then wash as normal.

Fabric Conditioner
If you are using wool, cotton and other natural fabrics, you should not need fabric conditioner (their purpose is really to reduce cling static in artificial fibres). If you feel that softer water will give your woollies a softer feel, try a little white vinegar (fragranced with essential oils). You can make your own fabric conditioner from 1 part white vinegar, 1 part bicarbonate of soda and 2 parts water. Use just as you would use manufactured fabric conditioner.

Floor Polish
Use good wax polish for wooden floors, which you should cherish and treat with respect. A good choice would be traditional beeswax. Lino and tiled floors will suffice with a mop over with warm soapy water.

Fly Killers
Avoid fly sprays. They are bad for our environment and bad for our health. Make sure that food is well covered or shut away and you will attract fewer flies in the first instance. Keep basil plants in the kitchen, flies hate them. They will also avoid orange and lemon peel, cloves and mint. You can make your own fly papers with brown paper strips and sticky sugar solution. Or use a non-toxic old fashioned fly swat.

Fruit Stains
A good way of dealing with the problem of fruit stains is to stretch the fabric and pour boiling water on it from a height (the safest place to do this is in the bath tub). You can also cover the stained area in dry borax and pour water through the fabric.

Furniture Polish
Use wax polish (beeswax is a good choice) but use it sparingly and only once or twice a year. Use a cloth dampened with water and vinegar to remove smudge marks in between polishes. Olive oil is another good option for wooden furniture that is badly stained. Just mix it up with a little vinegar and apply.

Glycerine
Glycerine is an inexpensive product available from any chemist. If you wipe around the inside of a fridge or freezer with glycerine on a rag, it will speed up the defrosting process next time. Glycerine is a useful stain remover for curry stains too. Rub some into the stain, leave to soak for an hour then wash as normal.

Grass Stains
Try rubbing with eucalyptus oil, and then washing as usual.

Hairbrushes
Use soda crystals or bicarbonate of soda to clean your bristle hairbrushes. Stand the brush in the solution so that only the bristles are immersed. To dry clean them, sprinkle them with flour, take a brush in each hand and rub the bristles against each other.

Hard Water Deposits
To remove deposits from vases, glasses and so on, fill with vinegar, leave for a time, then use a scouring pad to remove. Use the vinegar again for cleaning your toilet.

Ink Stains
Ink stains can be removed from cotton fabric by rubbing them with half a lemon. Just wash normally afterwards.

Ironing
If you need to descale your iron, use a strong vinegar solution. If you use distilled water it will not need descaling. If your iron feels sticky, rub a piece of soap wrapped in a hankie over the face a few times. Don't use aerosol spray starch for ironing. You can buy old fashioned starch that you mix up yourself. Put this in a plant spray and you have a perfectly acceptable substitute.

Kettles
To descale it well without buying commercial preparations, fill with 1 part water, 1 part vinegar and bring to the boil. Leave the kettle to cool and hey presto you will have a well descaled kettle.

Lemon Juice
Lemon juice is useful as a mild bleach. It can be used for cleaning copper and descaling pots and pans.

Mops
Use string mops rather than disposable paper types.

Onions
A great way of cleaning up steel and tin is to rub well with half an onion. Leave on for a day or two, then polish up with an old rag.

Oven Cleaners
Once again do not succumb to using aerosol spray oven cleaners. Make a paste from baking soda and water, or a solution of washing up liquid and borax. This can be made into a spray by using a plant spray as a container.

Pans
Soak burnt pans in salt water overnight and bring to the boil next day. The burnt offerings should then dissolve without too much problem. Avoid scouring wool or similar abrasives on stainless steel pans, it is said to release toxics into cooking afterwards.

Pewter
Wash pewter regularly in soap and water and then try polishing with a cabbage leaf (organic of course!).

Quarry Tiles
If you find you have whitish patches on your quarry tiled floor, brush them with a weak vinegar and water solution. Repeat if necessary.

Rugs
Collect old tea leaves in a jar with a little salt. Use this mixture to clean rugs and carpets in dusty rooms. Just sprinkle it on the carpets and then hoover it up. It attracts the dust remarkably well.

To clean oriental rugs, lie them face down on damp grass and then walk on them. To complete the operation hang the carpets over your washing line and give them a good beating to bring them up really well.

Rust
Lemon juice and salt will usually do the trick to remove rust stains.

Salt
Salt can be used as a mild disinfectant and makes an effective scouring powder. A handful of salt and a kettleful of boiling water, will also keep your drains clear, if used regularly.

Salt can be used to dry out damp cellars. If you place half a pound of salt in four tins and stand them in the four corners of the cellar you will discover that the salt absorbs the moisture, leaving the cellar drier.

Scourers
Try to use steel wool instead of scouring pads. It enables you to use as much or as little as you require, because you can cut off the wool as you require it. You can also choose your own soap, instead of using the soap in the middle of the pad.

Silver
If you soak silver articles in a saucepan containing hot water, a small piece of aluminium foil (or a couple of milk bottle tops) and tablespoon of washing soda, you will find that your silver will come up bright and sparkling. Never store silver plate in newspaper, as the printers ink could damage the silver plate.

Soap
Try to use a plain mild soap without deodorant or fragrance. If you buy the expensive fruit scented, glycerine soaps, use for a month or so as a drawer freshener. It will make your socks and underwear smell wonderful. Save all the end bits of soap in a soap jar, and add the occasional teaspoon of glycerine. When it is full, pour on a little boiling water and you will be left with a soft, jelly-like soap substance. This can be used for all manner of things around the house, including washing dishes, or – in a diluted form – to destroy aphids in the garden.

Spiders
Do not stamp on spiders if you see them around the house. They are important guests and should be welcomed. They feed on insects that carry disease, such as flies. They are also an important food source for birds. Learn to love and value this little cleaner, an important member of our ecosystem.

Talcum Powder
Make your own talcum powder from cornflour or arrowroot, fragranced, if you wish, with dried herbs, spices or dried flowers. Much commercially produced talc is mined in areas where there is asbestos naturally present, and could therefore present a health risk.

Tea Leaves
If you pour boiling water on used tea leaves, and leave them to stand for an hour, the liquid may be used for cleaning glass, furniture and lino.

Thermos Flasks
These can be cleaned effectively with bicarbonate of soda, which will also deal with stale smells. Two sugar lumps in the bottom of a thermos standing empty, will also prevent stale smells from developing.

Toilets
Avoid chemically-based commercial toilet cleaners and bleaches if at all possible. They delay the natural process of decomposition of excrement (a particular problem when untreated sewage is pumped into the sea). They can also be

a health hazard, especially with small children in the house, since they seem to have a nose for sniffing out hazardous substances! Far better to use vinegar, or vinegar-based commercial products, to remove stains, and a mild borax solution to disinfect.

Toothpaste
A good deal of controversy surrounds the toothpaste manufacturing industry. Greenpeace have campaigned to ban the production of toothpaste containing titanium dioxide (a whitener) because of the acidic pollution its discharge has caused in river systems. Other ingredients found include ammonia, formaldehyde, ethanol and saccharin. The natural varieties available from health food shops do not contain these ingredients, and invariably taste much better. You could even make your own toothpaste from bicarbonate of soda mixed with a few drops of oil of peppermint.

Vacuum Freshener
Pop a pad of cotton wool soaked in your favourite scented aromatic oil in the bag of your vacuum cleaner, and you will have a pleasant scented room when you have finished vacuuming.

Vinegar
As well as being a good toilet cleaner, white distilled vinegar makes an excellent smear-free window cleaner. A simple solution of vinegar and water, placed in a spray container will clean windows just as well as commercially produced window cleaners (which often contain ammonia). Be prepared to use a little more elbow grease though. This 50/50 water-vinegar solution will also clean tiles in your kitchen and bathroom, and will descale your kettle and remove stains from your teapot.

Washing Powder
Try airing clothes to get an extra day's wear out of them. This will help cut down on the amount of washing powder you use and help to save energy in your home. Use water softener to reduce the amount of detergent required, if you live in a hard water area. Use a washing powder that takes environmental factors into consideration, such as Ecover or Ark.

Cleaning

Many supermarket chains have now developed their own 'green range' of cleaning products, including washing powders. Try them all and find the one that suits you best.

Washing-Up Liquid
The trick with washing-up liquid is to use as little as possible to minimize environmental damage. Ecover and Ark both produce pleasant scented liquids that are less harmful to the environment, or you could produce your own from your soap jar. A few tablespoons of white distilled vinegar will help to cut grease effectively and you will need less washing up liquid as a result.

Water Softener
A good water softener will enable you to reduce the amount of washing powder you normally use. Washing soda is the simplest and least expensive form of water softener you will be able to purchase. It will also help you to deal with really stubborn stains.

White Linen
To whiten your bed linen, napkins and so on, give them a really hot wash, and then spread them out in the garden and allow the sun to bleach them naturally.

Wine Stains
To remove wine stains, squirt with a soda syphon. This will normally deal with the stain successfully. With stubborn red wine stains, place a handful of salt on the stain, and leave it to stand overnight. Then pop clothes into the wash as normal.

Zinc
To clean zinc, wipe with vinegar, rinse well, dry and polish. Or rub well with bicarbonate of soda on a damp cloth, dry, and polish up.

Shopping List for Basic Cleaning Materials

Large tin baking powder or bicarbonate of soda

Bottle white vinegar

Environmentally safe washing up liquid

Soda crystals or borax

Old towelling off-cuts for polishing and cleaning

Beeswax polish or olive oil

Lemon juice

Steeel wool scourers

Laundry

Environmental washing powder

Separate bleach box

Salt (for stubborn stains)

Large bar of soap

Eucalyptus oil

6 Food and Drink

Introduction

Every decision we make about the food we eat is a vote for the kind of world that we want to live in.

Frances Moore Lappe, 1971

Food is the most basic and essential part of our daily routines. We rely on it to give us energy and sustenance, allow us to grow and keep us healthy. Most of us take food for granted, however. How it is grown, harvested, manufactured and produced means little to us as long as we can eat as much as we think we need. But recent scandals and new information concerning the safety and quality of our food have brought new worries.

As consumer confidence in the safety of food has dropped, the active consumer has begun to question the food industry more closely. With scandals like listeria in cooked and chilled foods, salmonella in eggs and meats, BSE and dioxins in milk, as well as use of excessive and dangerous food additives, concerned consumers have to be fully aware of information and issues that may adversely affect their health and the environment.

There is a clearly established link between the food that we eat and our health. Doctors, nutritionists, food experts, dietitians and politicians all agree that the health of our bodies and the health of the planet are directly connected.

Food – a Global Issue

Back in 1971, Frances Moore Lappe, author of *Diet for a Small Planet* warned us of the environmental effects of our eating habits. She called for 'food consciousness', explaining that we are directly responsible for degrading the soil and depleting precious resources by eating 'high' on the food chain. By this, Moore Lappe means that in order to supply us with high-quality protein foods like meats, large quantities of grains and other important foods are fed to animals, to provide us with protein which could easily come from other sources. One half of the harvested agricultural land in the United States is planted with feeding crops and over 78 per cent of the grain is fed to animals. It takes 16 pounds of grain and soyabean to feed a cow in order to produce just one pound of beef. Twenty vegetarians could be fed using the amount of land needed to graze and feed just one meat eating person. Moore Lappe's point is simple: we eat too much meat.

When we consider that an estimated 60 million human beings literally starve to death each year from hunger and hunger-related diseases, and millions are at the the point of starvation, a recalculation of our diet seems to be a vital step towards human and ecological equity. Other concerns have been voiced, over the excessive use of antibiotics and additives fed to our livestock. The animals also produce millions of tonnes of manure, which is usually treated as a waste product and washed away into rivers and lakes, changing the ecology of our waterways. Factory farming, where cattle, chickens, pigs and other livestock are squashed together in tiny, cruel homes, often unable to move and rarely able to look after their offspring, has caused an outcry from food and animal lovers alike.

The Ethics of Food
Some argue that the quality of our meat is directly connected to the fear and terror that the animals go through just before death. Intensively reared chickens and turkeys are genetically manipulated to make sure they grow with as much edible meat as possible, and the average chicken lives in semi-darkness in a vast hangar with tens of thousands of other chickens, often unable to move or see.

Slaughtering of both chickens and other livestock is usually done in slaughter houses. The animals are stunned with an electrical charge and then their throats are slit. One report in the United Kingdom by a leading consumer writer claims that 90 per cent of our slaughterhouses are contaminated and filthy.

Some companies are making huge changes, however. Organically farmed and humanely reared meat is available for those who look for it, and the vegetarian trend has gained acceptance as more people are turning to fully or partially vegetarian diets. Major supermarkets in the United Kingdom have begun to stock organic meats in some stores, and if the consumer likes the product enough the trend will hopefully catch on.

Additives and Pollutants
E numbers, the number given to a particular additive or chemical supplement to foodstuffs, have caused so much concern in the last fifteen years that manufacturers are now proudly telling us what *isn't* in our food. The average UK consumer eats about 3.5 kilograms of food additives each year, mostly through additives that are largely hidden. Some are not even mentioned on the labelling of our food, and continuous campaigns fight for the legal right for consumers to know exactly what their food contains. There may be about 6,000 different food flavourings being used on our food, but we do not have the right to know who produces them or why they are used. They are often used by the industry to encourage us to buy food made to look more appealing than it really is.

Pesticides are another source of concern . Nitrate run-offs in farm slurry, and from other agricultural practices, do great damage to the ecosystem. Spraying large quantities of pesticides and herbicides over crops is also very damaging. Pesticides are designed to kill, and they very often kill every creature that they come in contact with.

With so many worries attatched to the food we put into our mouths, it's little wonder that people tend to shut off . But food is the basic stuff of life. We need it to survive and if we do not fight for it to be clean, healthy and wholesome we are indirectly and directly helping ourselves to poorer health and a worse standard of living.

Directory

Additives
A food additive is a substance in food which has little nutritional value and is present to increase consumer acceptabili-

ty by enhancing the appearance, taste, colour, or shelf-life. There are an estimated 10,000 additives presently being used, many of which are synthetic chemicals that affect our health and well-being (many cause hyperactivity in children). Try to avoid buying overprocessed foods. Do your own cooking from fresh natural products. Check your food for additives on the manufacturers labelling, and arm yourself with a manual on food additives.

Alcohol
Alcohol, taken in moderation, has been proved to have beneficial effects on our health. It reduces stress levels and can have a relaxing effect when taken in small quantities. However we really should try to avoid alcoholic drinks that are full of chemicals. Organic wines are now readily available at most wine retailers, and CAMRA (Campaign for Real Ale) have been campaigning for many years to establish the sale of beer free from additives. If you have difficulty obtaining organic wine, you could always make your own from organically grown fruit and vegetables.

Bottling
Bottling food is an inexpensive, easy and healthy way of preserving. Fruit can be bottled in water, which is a healthier alternative to the traditional sugary syrup. You can prepare fruit for bottling in the oven or in a pressure cooker. Use proper preserving jars (like Kilner) for the best results, with rubber sealing bands and metal clips. For special occasions, you could substitute brandy or another spirit for water.

Bulk-Buying
Buying your food in bulk makes sense all round. It saves money, time, and energy. It also cuts down on the amount of shopping trips you have to make, so you will not be encouraged to make costly impulse purchases, and you will help the environment by cutting down on transportation.

Cash Crops
Cash crops are crops grown by poor countries to sell to richer nations to pay off high interest rates. They include maize, soya, cotton and sugar cane. These countries could easily produce enough food to feed their own population, but as a

result of their governments' debts, their own people live in poverty and face starvation. To help solve this problem we can buy tea, coffee, dried spices etc. from companies like Traidcraft, who ensure that the producers receive a fair price for their goods.

Chocolate

The chocolate market in the United Kingdom is worth £2.4 billion annually. We are a nation of chocoholics. The consumption of chocolate is not just of concern to dentists and nutritionists; environmentalists are extremely worried about the highly dangerous pesticides used to produce the cocoa bean. Try carob for a less processed healthier alternative and attempt to cut down on chocolate generally.

Coffee

Like cocoa crops, coffee bean plantations are sprayed with highly toxic pesticides. Organic coffee is available, but it does not take away the health risks associated with caffeine, such as high blood pressure and heart disease. Decaffeinated coffee does not provide a solution either, because the process used to extract caffeine uses chemicals and is environmentally suspect. Try to cut down on the amount of coffee that you drink, and do not use bleached paper filters to make filter coffee. They contain dioxins which are highly toxic. Use washable, reusable filters.

Convenience Foods

It is all too easy to nip into the supermarket on the way home from work and grab a ready made chilled meal to pop into the microwave for supper. It is, however, perfectly possible to cook quick, tasty and nutritious meals at a fraction of the cost without the problems of overpackaging, overprocessing and using a microwave oven. The greatest friend you can have in your kitchen to help you produce speedy meals is a pressure cooker. They cook brown rice and substantial lentil soup in only four minutes, and preserve most of the nutrients.

Cooperatives

Start up a food cooperative in your neighbourhood. You share the cost of bulk buying your food, share the responsibility of buying it and share the distribution of it. If there are twenty people involved in your cooperative, each member will only have to do the purchasing about three times a year. It is a marvellous way to ensure that you get fresh produce at the right price, and is an excellent way of networking in your locality.

Dieting

Arabella Melville, author of *Persistent Fat and How to Lose It*, claims that dieting releases stores of body fat to use for calories as a substitute for food. This has the effect of releasing harmful substances stored in fat deposits into the blood stream, such as dioxins. This is particularly true of crash diets, or rapid weight loss diets. If you do have a weight problem, try to lose weight sensibly over a long period of time to minimize the risk of overloading your system with a sudden release of toxins.

Eating Out

It is becoming easier to visit restaurants that offer good wholesome food at a reasonable price. Ethnic restaurants are often a good choice as they usually cook their meals on the premises from fresh produce. There are also many good vegetarian restaurants around, and many more restaurants now offer a vegetarian selection (not just vegetable lasagne, as was once the case).

Eggs

The average person in the United Kingdom consumes 200 eggs every year. Very few of these are free range. Although even free range may mean being cooped up in a barn with thousands of other hens, and almost as little space as battery hens. If you do have access to a producer of free range eggs, do visit and check for yourself the conditions before you buy. Genuine free range eggs are more expensive, but far more humane, and they taste infinitely better too.

Fast Food

In the United Kingdom alone the fast food market, which includes burgers, hot dogs and pizzas is worth £1.5 billion. Fast food is not nutritionally well-balanced. It is low in proteins and vitamins and high in fat, sugar and salt. It also contains an abundance of chemical additives. The industry uses vast amounts of meat. Hamburger chains have been responsible for deforestation of enormous tracts of rainforest, to create ranches to raise beef cattle for the burger market in the United States. After ten years this land is reduced to dust and is abandoned in search of new pastures, leaving environmental destruction behind. The fast-food industry is also responsible for massive amounts of litter, much of which is non-biodegradable. Fast food may be cheap and convenient, but at such a massive cost to our health and environment, it cannot be justified.

Fat

We eat an average of 135g of fat every day in this country, 27 per cent from fat, 30 per cent from dairy products. Experts estimate that we could cut our fat intake by a third without too much difficulty, and as a result decrease our risk of developing heart disease, and certain cancers. One of the most immediate ways would be to cut down on our meat intake. If we all followed this advice, it would have the added environmental benefit of reducing intensive livestock farming, which would improve the land and the lot of the animals involved.

Fish

Salmon has become widely available over the past few years as a result of fish farming in this country. The problem is that the farmed salmon lose their natural pink hue. To counteract this problem, the fish farmers add a synthetic food colourant to the salmons' diet to make them look pinker. The colouring is canthaxanthin (E161). The use of this colourant has led to much controversy about its possible links with cancer. As a result, certain supermarkets have removed farmed salmon from their shelves.

Food Poisoning

According to the London Food Commission, bacterial food poisoning cases have increased by 62 per cent over the last few years. Many of these cases are directly attributable to the food manufacturing industry. This has led to a call for tighter reins on the food manufacturing industry as a whole. The problem is that at the moment, we have a Ministry of

Agriculture, Food & Fisheries that has the interests of both producers and consumers at stake. Ideally we should have a separate Ministry of Food, which makes the needs of the customer a priority.

Gelatine
Many people are not aware that gelatine contains the leftover gristle and bone of animals, and would be appalled to know that many of the products that they buy contain it. Check jellies, sweets, biscuits, pies, even vitamin capsules! Wholefood shops do stock alternatives, such as agar jelly.

Herbs
Herbs add variety and subtlety to many dishes and can change a simple casserole into a culinary delight. The best way to use herbs is fresh from the garden. Most culinary herbs are very easy to grow and need little tending. Of course, it is not always possible to use fresh herbs, so it might be sensible to keep a supply of dried herbs for use out of season. To dry herbs, you should pick them when they are young and tender: early to mid summer is the ideal time. They can be dried in the airing cupboard, or a warm oven, or even in a warm airy garden shed. You will only need half the amount of dried herb as they are more potent than their fresh counterparts.

Honey
Honey is a natural sweetener and a healthier alternative to refined sugars. It is a delicious replacement in drinks and in fruit purees. Try to buy locally produced honey, from a local beekeeper instead of imported honey. Think of all the fuel that would be saved if we all bought local produce when able.

Irradiation
In August 1990, MAFF (Ministry of Agriculture, Food and Fisheries) issued proposals for regulations on irradiation of food for public consumption. If accepted, these will end the ban on irradiation that has been effective since 1967. Irradiation of food for public consumption is an extremely worrying proposition. It would make it possible for manufacturers to sell low quality food that would last longer on the supermarket shelf and still appear fresh and wholesome.

There is fear that some forms of bacteria (particularly botulism) could survive irradiation. It would introduce the need for new types of additives to deal with the effects of treatment. It would also offer health risks to those working within the industry with radioactive materials. At present most supermarket chains have claimed that they will not stock irradiated food even if it is introduced by legislation (with the exception of Sainsburys). And the government claim that all irradiated food will have to be clearly marked. It still remains a frightening development in the food manufacturing industry, and one consumers should really take issue about.

Labelling

Although we have far more labelling on food products than a few years ago, much of the information on offer falls short of what we really ought to know. The fact is that most labelling is produced at the whim of the manufacturers, not by legislation. Consequently, we are told what they want us to know in order to sell us their product. Negative information will be difficult to find, whereas strong selling points, such as no additives or preservatives will be emblazoned across the front of the product. This is particularly true of so-called green products, where the environmental hype hides the truth about the reality of the product. Instead of having to take a degree in food chemistry, avoid the problem where possible by buying simple and natural wholefoods that do not need further explanation.

Markets

Markets can be environmentally friendly places to shop. You can buy a good range of foods that will not be overpackaged or overpriced. Some markets have organic fruit and vegetable sellers, and some now have good wholefood stalls. All in all though, markets are a positive shopping experience, encouraging small businesses in the world of mega-retailers, and encouraging local shopping.

Meat

The cost of eating meat is extremely high as far as our planet is concerned. It costs the lives of millions of animals and uses land for raising cattle to feed a small proportion of the world's population. If the land were given over to grain production, we could solve much of the world's food problem quite easily. There is the ethical matter of factory farming of livestock, which means unneccessary suffering for millions of creatures every year. Intensive rearing also means the liberal use of antibiotics, hormones and other growth promoters. It would also be beneficial to our health to eat less meat by reducing our fat (and hormone) intake.

Milk

Milk and milk-based products are staple foods in the Western diet. Milk is also one of the most allergenic foods available and causes many problems, especially in children. Clinical tests have shown that milk is implicated in over 70 per cent of allergic skin reactions and almost 90 per cent of asthma and hayfever cases. The whole business of milk production is also open to examination. Producing milk involves depriving lactating cows of their young, who are then often taken off to be used for veal. Cows are also given various hormones and drugs to improve production levels. These are then passed to us through our daily pint. Once again, less milk consumption would mean less demand on precious land which could be used to produce grain. If you must drink milk, for your health's sake use organic or at least semi-skimmed milk and milk products.

Nitrates

Nitrates are used primarily by the farming industry as fertilizers. Nitrates themselves have a low toxicity, but in the body they form nitrites, which can poison the blood and starve the body of oxygen. Babies are at risk from 'blue baby syndrome' and babies in some areas in Norfolk are given bottled water because the authorities cannot guarantee the safety of the tap water. Agricultural run-off is the largest cause.

Olive oil

Heart disease and obesity represent a major killer in Britain, and are related to the amount of fat we have in our diet. One estimate reckons that we can cut out 25 per cent of the fat that we generally eat without any negative effects, in fact nutritionists argue strongly that, combined with sugars and carbohydrates, fats are a serious health risk. Olive oil is the one oil that is worth eating. Try cutting out as many oils as you can, especially the ones hidden in cheap, convenience foods like cakes and biscuits, and swopping frying oils for pure, cold pressed olive oil.

Organic Produce

Organic food is now widely available in wholefood stores, supermarkets and even in some restaurants. Organic food is produced without artificial fertilizers, chemicals or hormones. Soil is usually left for a few years so that the residues of chemicals are completely gone before the Soil Association will test and award food products their organic food symbol. One of the main reasons that people buy organic food is because it tastes better, even if it is expensive. At present about 55 per cent of our organic food is imported. Overpackaging of organic food has been scrutinized by environmentalists recently, and supermarkets have found pressure on them to reduce the amount of wasteful packaging involved.

Packaging

Around 70 per cent of the contents of your dustbin is packaging, including trays and boxes, plastic bags, jars and tins. The industry is worth over £5 billion each year. Packaging is made from paper, plastics and various metals like tin and aluminium. All these have an environmental impact and they are rarely recycled. You pay for all that rubbish around your food. Demand less, hand back overpackaging in the shop, buy in bulk, avoid buying products that are individually wrapped or contain various types of materials like plastics and paper, making them impossible to recycle.

Pulses

Pulses – foods like beans, lentils, chickpeas and split peas – are the backbone of any good wholefood diet. They are high in protein and offer a real alternative to meat, especially when combined with other protein-rich foods like tofu, nuts and dairy products. Pulses can be made into a wide variety of soups, stews, pies and similar dishes. They are exceptionally cheap and can be bought in bulk as they store well for long periods. Soaked overnight they will be easier to cook and they make a hearty and filling meal.

Raw Food

With the increase in pre-packed, convenience foods, we forget that we should balance our diets with raw food to increase and stimulate our digestion. Eating more raw food means that we eat more vitamins and minerals too, as these are often lost when the food is cooked and processed. Vegetables and fruit can be eaten raw, and so can nuts. Try cooking food in very small amounts of water to preserve the nutrients.

Seasonal Produce

Buying food in season is an important step for any green consumer. Produce is not only cheaper but it will also be fresher and is more likely to come from nearby sources, cutting down on the amount of pollution and transport. Avoid exotic fruits and similar produce that is imported by air.

Sprouts

Mung beans, alfalfa and chickpeas are just some of the seeds that can be 'grown' into live food which represents one of the most nutritious food sources that we have. Rinse the seeds and soak them for about 24 hours or until they split. Place them on a porous mesh or cotton. Water them at least twice a day and within a week you should have an incredibly cheap and simple food for salads, stir-fries and sandwiches.

Steaming

Steaming vegetables and food is an energy saving and nutritious way to cook. Food seems to taste much better if it is steamed and you can save energy by placing the steamer over another food while it is cooking.

Sugar

The average diet consists of large amounts of sugar; it is an important food source in its purest form, but for many years we have been eating highly processed sugars, which are dangerous to our health and have caused problems ranging from rotting teeth to allergies. Much of the 70 kg of sugar that the average British consumer eats each year is ingested through hidden sources when sugar is added to foods that you would not normally suspect of containing it. Try cutting down on your sugar intake and test various foods without it altogether.

Tea
Organically produced tea is now available in supermarkets and wholefood stores, and the best way to drink your tea of course, is without that supreme example of excess packaging – the tea bag! Try and drink tea from loose packets; you'll be surprised at how much money you will save.

Vegetables
Vegetables in their prime are fleshy or tough, colourful and extraordinary. They offer us a rich and abundant source of trace minerals and vitamins and roughage. Sadly however, the use of pesticides, fertilizers and chemicals has meant that what we once considered to be healthy produce may not be so good for us. Traces of dozens of pesticides have been found in most of our main vegetables. These may be dangerous to our health over long periods, although most of them can be washed off when we prepare the food. The damage to the environment is an important issue, however and depletion of minerals in the soil and chemical run-off into our water supply are major problems.

Vegetarianism
Cutting down on meat can be good for us; over one million people in Britain have opted for a complete vegetarian diet. This means that they avoid meat and meat-based products. Vegetarians appear to suffer less from heart disease, bowel cancer, gallstones and diabetes and, provided they eat plenty of wholefoods and pulses, they will find that they have a healthy diet. Vegans cut out all dairy products and animal derivatives, including produce like honey and clothing that comes from animals.

Wholefoods
Wholefoods are foods that are not manipulated, processed or treated. Less refined foodstuffs are important for any conscious green consumer who wants to eat low on the food chain. You can buy organic wholefoods in your local health or wholefood store and a stock of the basics will always stand you in good stead for emergencies. You should

include oats and barley, brown rice, wholemeal organic flours, pastas, lentils, beans and nuts. Buy them in bulk to save money and packaging and keep in jars and containers in your larder.

7 Equipment

Introduction

Consumers want the services that energy provides not energy itself
                                                            John Button

New technology has allowed us to create excellent, environmentally-sound equipment that can reduce the amount of energy we use, allowing us to prepare and cook our food with the minimum of fuss as well as reducing the amount of toxic chemicals, and other pollutants in the environment. Pressure cookers allow us to cook our food more quickly; solar appliances mean that we can do without batteries; and our refrigerators can now be manufactured without CFCs as well as being more energy-efficient and therefore using less electricity.

When we think of equipment for the home we usually think of domestic appliances, like televisions, tumble driers and microwaves. Other kitchen appliances like kettles, cookers, food mixers, coffee grinders, and electric carving knives are also gadgets that we often feel are a necessity in our lives. What is it that has made us think that we need all these things? Why do we crave, year after year, for the latest gadget or appliance that we simply could not do without before? And does it really save us time and energy?

Whose Convenience?

An interesting study by Christina Hardyment called *From Mangle to Microwave* suggests that there are elements of the history of household equipment that may not be as wonderful as they seem to be. Manufacturers seem to have two aims when they sell us new equipment: the appliance must last just long enough for you to have forgotten when you bought it, so that you will go back and buy another (usually about one and a half to two years later); and the manufacturers must create ever-higher standards of cleanliness so that you always aspire to buying new things. Of course, some new equipment *is* created to be environmentally acceptable, when manufacturers consider the manufacturing process as carefully as possible in producing the product.

In 1990 over 95 per cent of the United Kingdom population owned or used a refrigerator in their homes. It is often the single biggest piece of equipment that a household will buy. For the last forty years we have begun to see that a refrigerator that chills food, keeps it fresher longer and stores cooked food is an important and cost effective means of good housekeeping. Mrs Beeton's laborious methods of keeping food chilled and fresh would be too much for even the most dedicated green homemaker today. Preserving fresh vegetables and fruit usually meant making them into jams (hence the name preserves) as they could not last longer than a few days. So the refrigerator has certainly made an important contribution to our lifestyles. Close on its heels comes the freezer. Some 79 per cent of United Kingdom households now own a freezer but unless you have a really large family or you grow your own food it is difficult to make them cost effective, and many nutritionists say that frozen food is just not as good for you as the real thing.

We cannot do away with our refrigerators now, but we can insist that they are produced to the best possible environmental standards so that our food stays as fresh as we need it to and we save as much energy as possible. The technology to reduce pollution by nearly 80 per cent already exists and many state-of-the art models are on sale in Europe and in the United Kingdom. Companies like Zanussi, who make models primarily for the German market, and Gram who produce for the Danish, seem to be more conscious of the issues. Their models were energy efficient before we in Britain even thought of it! The difference between a good energy consuming refrigerator and a bad one is phenomenal; you could save yourself over £100 in electricity costs each year alone by buying the best model. So always ask

Equipment

before you make your final purchase. Of course each pound you save in electricity bills will save carbon dioxide in the environment.

We need to remember, however, that refrigerators also use CFCs (chloroflorocarbons), which cause more environmental problems damaging the ozone layer. Companies are working round the clock to come up with a 100 per cent CFC-free model which may be available to everyone soon. You can already buy a low-CFC model from most reputable manufacturers. All the other electrical equipment that you use around the home increases the carbon dioxide in the global environment, not only when you are using the product but also when it is being manufactured. The carbon dioxide pollution from all our electrical equipment is a massive 42 million tonnes each year. Add to that the energy used in the manufacture and the extra pollution created from transporting the goods, sometimes halfway round the globe, and the ease and convenience that you imagined soon turns out to be a problem.

Many houses are now sold with a dishwasher and cooker already in place. The dishwasher is a good example of how a company can sell us a product that we do not always want. Around 1886, Josephine Cochrane designed and constructed a mechanical washing machine, not because she hated washing up – she seemed to have plenty of servants and maids to do that for her – but because she was fed up of these servants continuously breaking her best crockery. The invention caught on with the hotel trade but curiously enough when women were questioned in 1915 as to their household chores and preferences the majority said that they enjoyed washing up at the end of a hard day! The dishwasher then was sold to us as a machine that could use hotter water than our hands to get our dishes cleaner. There is no proof that they get our dishes cleaner but we do know that they are only energy efficient if they are used when absolutely full, selecting the lowest temperature and washing large items by hand.

Gadgets that can do an innumerable number of jobs seem to be popping up everywhere, just watch your television as Christmas approaches, advertisements for new household machines that clean everything from the cooker to the cat seem to be the norm! But a greener lifestyle means serious thought as to whether we really need them or not.

One of the gadgets that many feel is useful and acceptable in the green home is the food processor. Using only 3kw of electricity per year, the food mixer can pulp vegetables that we might otherwise buy in cartons, it can mix and whip up desserts and pies that we might otherwise be tempted to buy in overpackaged preprepared, additive-laden alternatives. Ironically, the electric version that we know today was first invented in 1922 in a joint venture by Fred Waring and Stephen Plplawski. They wanted to sell machines to make their favourite drinks, including daiquiris and milk shakes. It was marketed and sold mainly to bartenders for the first twenty years of its life and only became a household item in the 1950s, after a massive 'reduction campaign' to persuade housewives that they needed the handy device. The food processor as we know it, which can mix, mince, squeeze, slice, and shred, began in the United Kingdom in 1947 and has been adopted by households all over the world since.

Other household equipment may not be so useful however. There cannot be much justification for the electric carving knife, useful if you need to cut dozens of chickens and roasts perhaps but a waste of resources and energy for most of us. Electric can openers, and electric pasta makers fit into the same category.

One of the main arguments used by advertisers is that the gadgets will save women more time in the home. However, many studies say that women spend exactly the same amount of time on household chores as they used to fifty years ago. The use of gadgets and equipment may have made our lives more sophisticated, but it has ensured that we need to clean more fiddly items, replace equipment more often and of course, spend more time buying.

It is frustrating to think that the equipment that we buy today may only last a few years. The technology to make more durable goods does exist – in fact goods made fifty years ago are often still in working order. Those made today are built with costs in mind. They are effective but less durable. When something breaks it is often as economic to buy a new, up-to-date appliance as to have the old one mended. This approach is perpetuated by the manufacturers. This wasteful consumerism may have given us short-term employment but it has certainly affected our long-term ecological stability. To continue producing often useless gadgets with no concern for the raw materials and the energy use as well as the disposal problem can only be considered as unacceptable in a greener future.

People are now looking carefully at long-term, sustainable solutions and alternatives. Take batteries for example. New EC regulations have meant that cadmium and mercury have had to be removed from the process, creating less toxic pollution in the manufacturing industry. This has had a beneficial effect, promoting the sale of rechargeable batteries and helping people to understand the ecological reasons why we should avoid battery-operated equipment whenever possible. The battery itself will never by truly 'green' as it takes up to 90 times the amount of energy that you get from the product to produce it in the first place.

This in turn has helped the promotion of solar-powered options, made from silicon, phosphorus and boron. They never need to be renewed, end up saving resources and the consumer money in the process.

So, how do we assess whether the equipment we need is important, for our needs and manufactured in the most ecological way? Here is a list of guide-lines to help you when choosing equipment for your house.

Guidelines for Buying Equipment

1. Check the necessity of the equipment. Is it something you only need to use a few times? Can you borrow it from a neighbour or friend? Could a group of you pool resources and buy one?

2. Can you repair it yourself when necessary? Are there repair services with the contract? Is the company committed to repairing?

3. Is it produced by a company committed to paying workers a fair and equitable salary? Has it been made as locally as possible, to cut down on importation costs and pollution?

4. Does the equipment do exactly the job you need it for? What are the 'extras'? Are they gadgets designed

> to make you buy it or really useful additions to the design?
> 5. Does it use as little energy as possible? Is it possible to buy a model that is not electrically operated?
> 6. Is the equipment made from non-toxic materials?
> 7. Is the design suitable for everybody? Can disabled people and the elderly use it?

Directory

Batteries
Batteries can contain lead, cadmium, and mercury, all dangerous pollutants. They require ninety times the energy they give out to produce. If you must use household gadgets that require batteries, use the cadmium and lead-free varieties, or better still, buy a battery-charger and a stock of rechargable batteries, and save money in the long run.

Beds and Bedroom Furniture
Try to choose a bed that uses wood from a sustainable European source rather than a tropical hardwood. Some companies make their beds from European pine treated with natural organic oils and waxes. The Friends of the Earth *Good Wood Guide* will advise you on the best choices when purchasing wooden furniture (see page 148).

Chlorofluorocarbons
More commonly known as CFCs, these can create a big problem for those of us wishing to dispose of a fridge or freezer. Check with your local authority – some now offer a CFC recycling service. Some retailers will offer this facility if you buy a new model from them.

Clocks
The ticking of a clock can be one of the most reassuring and comforting of sounds to the human ear. Try to buy a good quality hand-crafted clock with a manual wind-up mechanism, rather than a battery operated or electric clock. If we

only consumed energy where it was strictly necessary, we could conserve a great deal. Even tiny savings are worth considering.

Cookers

Although electric cookers are cheaper to buy than gas models, gas is over 40 per cent more energy efficient for cooking than electricity, so use it for cooking if you are able to choose. If you are using the oven, use it for batch baking so it is used to maximum effect. Some cookers have a smaller second oven, but they are not necessarily quicker to heat up or cheaper to use than the main oven. However, the half grills and dual-element rings on the hob *are* energy efficient, and it would make sense to use them. Consider using the hob or the grill instead of the oven when you are cooking smaller portions. Try not to site your cooker against an outside wall as it will lose heat and will not be as efficient.

Dishwashers

Ensure that you only use your dishwasher when it is fully loaded to make it energy efficient, and if you have an economy programme, use it! A fully loaded 'normal' programme will use twice as much energy as your economy programme. A good tip to consider is to turn your dishwasher off before it reaches the dry cycle, and open the door. The dishes will be extremely hot and will dry themselves in no time. Ensure that your dishwasher is well insulated and it will be more energy efficient.

Dustbins

Try and reduce the use of your dustbin by recycling everything that you possibly can. Find out where the local recycling centres or skips are and use them. Food waste and other organic waste can be used for compost. Try putting a couple of handfuls of straw in metal dustbins to absorb grease and unpleasant smells, or deodorize plastic bins with a sprinkling of bicarbonate of soda.

Electric Blankets

A hot water bottle will warm a cold bed as efficiently as an electric blanket. Recent studies have linked the electromagnetic radiation from electric blankets with cancer and with miscarriages. Flannelette sheets will keep you warm and cosy in bed in winter and will not pose a threat to your health.

Electric Toothbrushes
It may seem trivial to criticize the moderate energy consumption of the electric toothbrush, but in many ways it represents the ultimate in unnecessary gadgets. A good strong traditional toothbrush, and a good supply of dental floss will clean your teeth perfectly well and give your arm and wrist a workout at the same time!

Food Containers
Use sturdy food containers for storage of food in your fridge. Never use disposable cling film or plastic packaging if you can possibly avoid it. Shop around for 'cadmium free' plastic containers. Cadmium is a toxic heavy metal and once released into the environment it becomes a threat to health. Boots and other major retailers usually stock a good supply of cadmium-free plastic containers.

Food Processors
When you consider that in terms of electricity it costs 6.5p to blend 500 pints of soup in a blender, or to mix a cake a week for a year, it is clear that a food processor is an extremely economical piece of electrical equipment. It can also save you time and money by helping you to prepare nutritious and delicious food at a fraction of the price of ready made supermarket fare.

Freezers
If you own a freezer, try to keep it well stocked with seasonal fruit and vegetables (preferably organic) and batch-cooked foods at all times to make it energy efficient. If you are contemplating buying a new freezer, look carefully at the energy efficient models, and those that have a low CFC content. Before buying a freezer however, think carefully about whether you really do need one. They use twice as much energy as a fridge, and are often left standing half empty. Wouldn't the freezer compartment of your fridge serve you just as well?

Gadgets
Do we really need an electric knife when a well sharpened

carving knife will do the same job, or a battery operated mini vacuum cleaner, when a dustpan and brush would suffice? Try to evaluate what gadgets you have in your home, and how useful they really are.

Garden Tools
A sound ecological idea which works well with garden tools (and DIY equipment) is a 'tool pool'. Get together with a circle of friends and neighbours and share your resources. It is much more economical than tool hire. The next obvious step is to share your skills as well as your equipment.

Glasses
There are light green heavy glasses and storage jars now available made from recycled glass. They are functional, hardwearing and are very price-competitive.

Halon Fire Extinguishers
Do try to purchase a model which is halon free, or use a good quality fire blanket. Halon fire extinguishers contribute to ozone depletion because they contain CFCs, and are therefore not recommended for the green home.

Humidifiers
Many of us buy a humidifier to counteract the effects of overheating our homes. Try to limit the use of central heating to when it is really necessary, and you will find that you will have fewer sore throats as a result. Bowls of water and plenty of house plants will increase humidity effectively.

Ionizers
These could be a worthwhile investment, especially for sufferers of asthma and hayfever. The negative charge created by an ionizer gives the air that 'after the storm' quality, which can reduce tension and increase alertness.

Ironing
Try to iron as little as you can get away with. Who cares if your towels, sheets and underwear are not ironed! Use a heat-reflective ironing board cover to increase the efficiency of your iron. Try not to get the iron out for one item of clothing, resolve to make it a policy to iron in bulk.

Jars
Used jam jars and other jars can become a valuable form of household equipment. They can be used for general storage, to contain home made jams, chutneys and so on, and they can be used to sprout alfalfa, mustard and cress and other sprouting seeds and pulses.

Kettles
Try to use an electric kettle that switches off automatically when it comes to the boil, and preferably one that can boil as little as one cup of water. If you use a gas ring to heat your kettle, buy a whistling kettle to remind you when it has come to the boil. Remember to keep your kettle descaled to keep it functioning well.

Knives
Gradually accumulate a set of really good knives that are going to last for many years. Treat them with the respect they deserve, keep them well cleaned and sharpened, and you will never need to replace them.

Microwave Ovens
It is widely accepted that microwave ovens do use less energy than conventional cookers. However it is also recognized that only 35–40 per cent of the energy it uses is used to cook the food – the rest is wasted. In a government report published in 1989, over 30 per cent of ovens tested failed to meet the required temperature to kill bacteria such as salmonella and listeria. Microwaves also encourage us to buy overpackaged ready made-meals which are an expensive and wasteful way of feeding ourselves and our families.

Muslin Bags
Make a supply of simple muslin bags that you can take with you when you shop for fruit and vegetables: then you won't need to use wasteful plastic bags. You may get some strange looks at supermarket checkouts, but take heart and think of the environment.

Nursery Equipment
How many babies really care if their cot quilt and bumper are colour co-ordinated, or if they sleep in an expensive cot

or in a drawer? Be sensible, the equipment you are buying is usually a very short-term investment. Look in your local newspaper ads or in your newsagents' windows to find second-hand or recycled equipment, or ask friends and relatives if they can help. You'll be amazed at what good quality equipment can be picked up at a snip of the original price.

Party Equipment
When throwing a party, instead of using disposable tablewear find out if there is a local catering service who will hire you crockery, cutlery, table cloths, napkins and so on. Many off-licences will lend you glasses free of charge, if you buy your drinks from them. Alternatively ask around; you may find you can gather enough from friends and neighbours, and save yourself some money.

Pens
Ask for a good quality pen for Christmas or your birthday. Not one that uses disposable cartridges – one that you can refill with ink. This will save you using wasteful disposable biro type pens (which we buy in millions every year).

Percolators
If you must drink coffee, use a percolator or cafetière, instead of a coffee filter machine with its wasteful paper filters. Or buy a reusable cloth or metal filter. It may, however, be kinder to the environment and your health to consider giving up coffee altogether. Coffee plantations are causing considerable environmental damage, chiefly because of pesticides used. Also the negative long-term effects of coffee consumption, such as fatigue and high blood pressure, cannot be ignored.

Personal Stereos
It has been proved that all personal stereos are capable of an output the equivalent of a pneumatic drill, and consequently are capable of causing long-term hearing problems. Noise pollution is as great a problem as any other form of pollution, and must be kept in check. If you are a user, make sure you have the machine at a sensible volume. Personal stereos also cost the environment dearly in terms of the rate at which they use up batteries.

Pressure Cooker
Pressure cookers save time, energy, money and nutrients. Once mastered, they perform as well, it not better than, microwaves. (There is no reason why you shouldn't be able to produce a decent casserole in 12 minutes.)

Radios
Use a radio that runs from a mains supply and not a battery operated one. Even if you only listen for two hours a day, it is forty times more expensive to use batteries.

Rain Barrels
Why not consider using a rain barrel to collect precious rainwater? A large plastic dustbin will hold 15 to 25 gallons of rainwater, which could them be used to wash the car or water the garden.

Razors
Every day in the United States, 5 million disposable razors are binned. Because of the energy required to make them, they are a waste of the worlds resources, so don't use them. Don't use aerosol shaving foam either. It may be labelled 'ozone friendly', but the propellants which have replaced CFCs will be precursor greenhouse gases and are therefore unacceptable. Compromise and use a razor that only uses disposable blades and use old-fashioned shaving soap. Better still, use an electric razor – one of the few electrical gadgets that are an acceptable alternative because they use very little energy and will last you for many years.

Refrigerator
Keep a fridge thermometer in your fridge to monitor its efficiency. Remember to defrost your fridge regularly, as this will improve its performance and it will last longer. If buying a new model, look for low CFCs and energy efficiency. Take care to position your fridge away from heat sources such as cookers or direct sunlight, as this reduces its efficiency, and increases costs. Try to keep the door open for as short a time as possible (a very difficult lesson to teach children!).

Remote Controls
The best advice for remote controls is to get rid of them. They

make us lazy and use wasteful batteries. They also encourage us to leave televisions on standby, which means that our televisions are using a quarter of their power even though we are not watching them. According to Friends of the Earth, they are also generating an extra 2,000,000 tonnes of carbon dioxide every year , and costing us £12 million per year.

Saucepans

Don't waste heat by using a larger saucepan than you actually need. Use only enough water to cover vegetables and turn down the heat once the water has come to the boil. Always keep saucepan lids on to conserve energy and make sure that the lids do fit snugly. Use cast iron, enamelled or stainless steel saucepans and try to avoid aluminium. Recent research suggests a link between aluminium and development of Alzheimer's disease (pre-senile dementia).

Showers

For the amount of water that we use for the average soak in the bath, we could take five showers. The average filled bath can use as much as thirty three gallons of water! So please keep your long soaks for special occasions.

Steamers

Buy a two-or-three tiered steamer, and you will be able to cook a three course meal on one ring of your cooker. Buy a stainless steel or bamboo steamer for best results.

Sunbeds

Sunbeds only use the rays which tan your skin and cut out the rays occurring naturally in sunlight that burn and redden your skin. However, the tanning rays are probably as damaging as burning rays in the long run because they cause irreversible ageing of the skin. If you must use one, use an appropriate sunscreen, and never forget to wear the goggles provided to protect your eyes.

Teapots

Tea is still the most popular British drink. Every year we drink an average of 1,355 cups of tea. A reassuring fact, because tea on the whole has few chemical additives and is far healthier than coffee. What we are moving away from, however, is the use of the teapot. More and more tea manufacturers are producing 'one cup' teabags, which are an expensive and environmentally costly way of marketing a basic commodity. Stick to your teapot and use loose tea. That way you won't have to worry about the waste of paper or dioxins present in bleached paper tea bags.

Teflon-Coated Cooking Utensils
Teflon is a fluorine-based plastic which will gradually release its constituents. Eventually the surface will wear away and tiny particles will become embedded in the food you are cooking. Buy enamelled pots and pans or uncoated cast-iron pans.

Telephones
British Telecom recycle about 4.5 million telephones every year. Not just the electrical components – the plastic is melted down and reused as well. So don't just throw your telephone away, contact British Telecom.

Televisions
Electromagnetic radiation emitted from television screens is causing considerable concern. Low-radiation televisions have been developed by a Swedish Company, and are available in this country; but obviously if your current television is functioning well, you will not want to splash out and buy a new model immediately. In the meantime, try to sit feet away from the screen (particularly children) because electromagnetic radiation dissipates at this distance. The Swedish television is also low-energy, an important factor when we consider that Britain's televisions create 7 million tonnes of carbon dioxide, 10,000 tonnes of sulphur dioxide, 185,000 tonnes of nitrogen oxide and 830 cubic meters of radioactive waste every year.

Thermos Flasks
Use thermos flasks to store excess boiled water from the kettle. This makes economic sense, particularly if you filter your water.

Toasters
Using an electric toaster is a far more energy efficient way of

making toast than switching on the grill of your electric cooker. Alternatively if you have an open fire you could toast your bread and muffins the traditional way!

Tumble Driers
Tumble driers are very energy intensive, so use them only when absolutely necessary. Use the washing line during dry weather, and indoor drying racks or clothes horses during wet or cold weather. Remember to unfluff the filter in your tumble drier regularly as this will increase its efficiency. Another good tip is to turn the drier off halfway through the programme, and leave the clothes to dry in the warm machine. A gas tumble drier is now available at gas showrooms which claims to save on energy.

Vacuum Cleaners
Well-swept or vacuumed carpet will stay in good condition and will last much longer. It is worth considering whether the electrically-operated cleaner is always necessary. A dustpan and brush may suffice. Or it may be worth investing in one of the small manual carpet sweepers for daily use, with an overall vacuum once a week. Interestingly, cylinder vacuum cleaners use twice as much electricity as upright cleaners.

Video Recorders
If you have a video recorder and use it to watch films, hire them instead of buying them. Many libraries now hire out films and television programmes on video for a small charge. This is far more economical than buying them.

Washing Line
An often undervalued piece of kitchen equipment. Tumble dried clothes never have the freshly aired smell of line dried clothes.

Washing Machine
If buying a new washing machine shop around to buy the most energy efficient model available. Always wait until you have enough clothes to do a full wash, or use a half-load programme. Experiment with low temperature programmes. If possible, let the machine fill with hot water

from the gas heating system. Don't forget to use a washing powder that takes environmental factors into consideration, such as Ecover or Ark.

Waste Disposal Units
If you are thinking of installing a waste disposal unit in your kitchen, think again. It is an extremely wasteful and energy costly way of dealing with food waste. Compost all those peelings, skins, and leftovers, and use them in your garden.

Water Filters

Jug-type
It is very important to remember to change the disposable filters on these filters at least once a month. If not, they will release the pollutants, like heavy metals, back into the water that you have just filtered at some expense.

Plumbed-In Systems
Although they are more expensive than jug models initially, they do work out considerably cheaper in the long run and are much easier to use. Have your water analysed by a local water company before deciding on the type to buy. Different systems will suit the water in your area. Use a consumer publication like '*Which*' to help you make your decision.

Wicker Baskets
Keep a good supply of wicker baskets for storage. They will last for many years and come from a renewable source. Buy your baskets from Traidcraft or Oxfam, who sell many household items made by people in the Third World, and the profit goes back to help Third World communities.

8 Junk and Waste Disposal

Introduction

In a sustainable society, durability and recycling will replace planned obsolescence as the economy's organising principle and virgin materials will be seen not as a primary source of material but as a supplement to the existing stock.

Lester Brown

The average individual in the United Kingdom throws away one tonne of rubbish each year: plastics, metals, glass, paper, textiles and food wastes. Our rubbish mountain has grown as our consumption has grown. We 'need' more packaging round goods than ever before. We 'need' to be in fashion, so we buy more clothes than before; and we throw away furniture and goods so that we can buy the latest style, or because they have broken prematurely.

One extreme of the post-industrial society is the Japanese example. On the rubbish tips and landfills of Japan today, you can find discarded machines that may only need as little as a new plug or similar small thing to make them workable. Such is the fashion, however, that we discard items as 'junk' well before their useful life is over, and often buy another replacement just to keep up with current trends. Sophisticated technology can now produce smaller machines, micro chips, screens and equipment directed towards energy gobbling consumption.

Consumption

As the century of the consumer began we thought that we should replace personal service with self-service and fast service. This shift in consumer habits brought about a massive change, requiring large amounts of mechanization, production and packaging, which in turn created large mountains of waste. Our economic system requires us to produce more, if we are workers, and consume more, if we are buyers. Little emphasis is placed on renewing, recycling and reusing. We need to consume, we need things to burn out, to be discarded and wear out so that we can replace them all over again.

Of course, this sort of consumption creates waste and environmental problems in huge quantities. The production of packaging alone uses 6 per cent of the United Kingdom total energy bill, and even the industry itself agrees that only two-thirds of our packaging is for food protection and hygiene; the rest is created to sell or market a particular product. Many environmentalists believe that we could cut down on our packaging by more than half, thus saving energy, and reducing pollution in the form of carbon dioxide and other greenhouse gases. The packaging costs money, of course, and it is the consumer who pays the bill in higher prices for food and drink.

Dangers of Waste Treatment
In general, our waste collection system takes the problem away for us. We buy too much packaging in the supermarket on Saturday, discard garden debris on Sunday, and left over bottles, cans and glass during the week, and we expect the dustbin services to pick everything up and get rid of it for us. About 90 per cent of this rubbish is simply put into a large hole in the ground, called a landfill site. Useful, reusable material is mixed with plastics and mixed materials. The cost of landfilling has been kept artificially low, and only in recent years have we realized that the waste does not degrade easily or quickly – even for readily degradable products like food wastes – and some items, like plastic, will never degrade.

Some materials discarded on landfills are not so simple to categorize. Batteries may be extremely corrosive when they start to break down; partly-used cans of cleaning materials, paints and other products may contain heavy metals and other toxic ingredients; poisonous pesticides used in our gardens are dumped on sites. CFCs leak from old refrigerators, releasing their fumes and destroying the ozone layer.

The noxious chemical cocktail found in landfill sites can leak into our ground water supplies. Chemicals like solvents, mercury, cadmium and other toxins make up a growing portion of our groundwater, which we eventually drink. Methane is also a problem in landfill sites, causing explosions and adding further to the greenhouse effect.

Incineration is sometimes the preferred option for disposal, although only 8 per cent of our waste is incinerated at present in the United Kingdom. As a method of disposal, many environmentalists agree that it is a last-resort method of getting rid of waste.

By burning the rubbish you are effectively turning the mass of mixed rubbish into a smaller amount of poisonous ash and a toxic mixture of gases which you can't see. It is thus harder to control. Toxic sludge also results, which then has be disposed of in a landfill. Some groups campaign for our rubbish to be burned and the energy created to be used for everyone, but this has been vehemently argued against by groups like Greenpeace, who believe that a safe waste disposal policy could never include airborne pollutants, especially where toxic waste is involved.

Alternatives to Dumping
There is much that the ordinary person can do to reduce levels of waste and help to create alternative markets for products that may well be dumped otherwise. Clothes are a good example. Each year we throw away over a million tonnes of clothing in the United Kingdom. Most of that is wearable or could possibly be recycled. The charity Oxfam collects second-hand clothes and puts them to good use. They sell the valuable clothes and raise money for their overseas projects and help fund their offices and shops. The rest of the clothing is sent to a massive clothes recycling depot, where the fabrics are chopped up and manufactured into bedding, jackets and other hardwearing materials.

Attending and rooting through the assortment of regular jumble sales, car-boot sales and bring and buys each Saturday or Sunday can be a fun hobby and turn up some notable surprises. Be careful that you don't end up with a home full of someone else's junk, instead of getting exactly what you need. Often a good place to pick up furniture of any kind is in an auction hall. Check local papers or library noticeboards for details. Imagine the bargains that you could pick up and think of all the new furniture that you won't need to buy. You will often find a much more stylish piece by hunting around, instead of buying the mass-produced version in the current fashion.

Restoring and transforming furniture or rubbish of any kind takes a particular skill and eye, to spot the value and potential of a particular piece of so-called junk in the first instance, and then to carry restoration or adaptation work.

As recycling of any kind of material or junk is considered to be the most important first step on the way to using our resources more wisely and cautiously, a good idea is to support any potential recycling scheme in your area, including buying products themselves made from recycled materials.

In North Yorkshire, a company has just begun to sell top quality clothing made entirely from recycled cottons and wools; no more cotton has to be grown and no more pesticides have to be used. The sheep get a better life too – their coats aren't taken away quite so fast! One of the easiest products to find these days is recycled, unbleached paper. Try to persuade the place where you work to switch to recycled paper and start collections so that you stop disposing of all that computer paper and letter writing paper. Paper itself can be recycled up to eight times, but it is rarely even recycled once at present. Recycling is a modern day growth industry, the employment of the future, so support it whenever possible. Each time you recycle you save resources, not just the original resources of the tree or the mining of the metal, for example, but also energy, water and a host of other precious resources. In order to recycle paper we can cut the amount of air pollution by 74 per cent, water pollution by 35 per cent and the energy that it takes to produce new paper by 40 per cent. This means that we could have much cheaper product in the longer term too.

Some Tips for Recycling

1. Separate your rubbish so that it can be recycled.
2. Always opt for a reusable product instead of a dis-

posable one. For example: use cotton handkerchiefs instead of tissue; use milk from bottles instead of cartons.

3. Reuse paper as much as possible, and then recycle it.
4. Buy recycled paper.
5. Take cotton or string bags to the shops with you.
6. Use any product right until the end of its useful life.
7. Learn how to fix things, to make them last longer.
8. Take care of your clothes, and recycle those that really aren't any good any more.
9. Protect your furniture or carpets from wear and tear.
10. Don't be a hoarder.
11. Cut down on packaging at the shops, if it looks too much, hand it back in the store after you've bought it, as a protest.
12. Don't be a litter bug – don't create a bad environment, never drop litter and report any major littering, especially dumping to the police or your local authority.

Directory

Auctions
Auctions may be conducted from a private house where the entire contents are being sold off, or from a salesroom. Dealers have a well trained eye and they will not miss a bargain very often. However, if they consider the item in question to be merely an item of junk, they will lose interest and you may well pick it up at a bargain price. When attending auctions, make sure that you know exactly what it is that you are bidding for and how much you are willing to pay for it. Go along to the viewing session before hand and have a look at what is on offer. Take a measuring tape to make sure that you won't have problems transporting it or finding a suitable space at home.

Basic Tool Kit

As a keen junk restorer, you will need a basic tool kit to enable you to carry out your restoring and adaptations successfully. This should ideally include a rule, a measuring tape, spirit level, a panel saw, tenon saw and padsaw. You will also require a selection of nails and screws, screwdrivers, a hand-drill, and several different sized chisels. Last, but not least, a hammer, mallet and plane for finishing. You may even be able to pick up a number of these items at auctions, or jumble sales. You are now equipped to begin your transformations of furniture.

Books

All is not lost with damaged books. If the cover is damaged, you could try your hand at book binding, or if the book is potentially valuable you could contact the Institute of Paper Conservation who will be able to refer you to a suitable book restorer. If you have piles of old books at home give them to your local hospital or charity shop. In an ideal world, we should be looking towards sharing resources like books, and making greater use of our public libraries.

Boot-Sales

Car boot-sales have taken off in a massive way in the United Kingdom over the last few years. They are a great opportunity for picking up junk and for disposing of your own unwanted goods. The more established ones attract large numbers of stalls and punters, and reputable organizers will limit the stallholders to second-hand merchandise to prevent maket traders selling new goods. They are a wonderful opportunity to practise haggling too – well recommended for week-end entertainment. Check your local newspaper for dates and details of your nearest.

Bottle Banks

Every year in the United Kingdom we throw away 6 billion bottles and jars. At the moment only 14 per cent of glass is recycled, even though recycling reduces energy requirements considerably (every tonne of crushed glass saves the equivalent of 135 litres of oil). By 1992 there are plans to increase the number of bottle banks by 50 per cent. To be even more environmentally friendly, reuse glass as much as you can and campaign for the return of refundable bottles. If we can do it with milk bottles, there is no reason why we cannot do it with others. In Oregon in the United States, it is illegal to retail non-returnable bottles. And in Sweden a tax has been levied on non-returnable glass since 1970. So start protesting about your lack of local facilities today!

Cardboard Boxes

It is often said that children gain more pleasure from the boxes than the toys that they contain. There are a 1001 uses for cardboard boxes apart from the obvious one of storage. Boxes can become toy cars and trains. They can be made into wendy houses or castles or spaceships. You don't have to do anything, just hand the box to the children and leave it to their imagination.

Charity Shops

Charity shops work to overcome poverty, at home and overseas. They raise money by reselling and in some cases recycling goods donated by the public. You may not always find exactly what you are looking for, but the serendipity of the place makes it a pleasure to spend time simply browsing. Try to phone before making a special visit, as most charity shops are run on a voluntary basis and sometimes have restricted opening hours. They will accept clothing, wool, toys, books, jewellery, records, pictures, stamps and coins as well as many other items.

Clothes

In the United Kingdom each year we throw away £400 million worth of clothes. Many of these are in perfectly good condition and could be put to good use. You could organize clothes swops among friends and relatives. You could give children's clothes to hospitals or children's homes, if you have no one to hand down to. You can always recycle your own clothes to give them fresh appeal. Try unpicking woollens to knit into new designs. Add decorative patches to threadbare areas. You'll find that you can adapt many items of clothing in this manner.

Collages

Collages are a terrific opportunity to create works of art from junk. Use any waste materials such as fabric scraps, corks, milk bottle tops, buttons, beads, in fact anything at all to express the use of junk in a creative and inspired fashion.

Compost

We throw away millions of tonnes of organic waste every year, that could be used to make compost. At the same time we are digging up peat bogs in Scotland and Ireland, to the detriment of the local environment, to do the job that composting could do. Vegetable peelings, orange skins, clippings, cotton wool, human hair, house dust and tea leaves are just a few of the things we can use for compost. You can build your own compost container with relatively little effort, or you can buy a plastic or wooden model from most garden centres. The National Centre for Organic Gardening has produced some good leaflets on this subject.

Dustbins

If we are to become successful recyclers in the home, we must make it easier for ourselves by separating our waste effectively. In Germany, it is quite common for each house to have colour-coded bins: blue for paper, yellow for metal, and so on. These are collected individually by refuse collectors. In this country, in Sheffield, on a simpler scale, the blue box collection means that all recyclables such as metal, paper and glass are collected in a blue box once a week. Even if it simply means having a box for your newspapers, bins for glass and aluminium, and a bucket for your compost heap, it pays to be organized. It really does take little effort to separate, once you get into the swing of it.

Electrical Goods

It is all too easy to dispose of your present washing machine because a new laser model that does everything except bath the baby has come onto the market. Electrical manufacturers and their advertising agencies have sold us the myth. There are, however, small firms and cooperatives around the country who do not accept that reasoning. They often train people to repair electrical equipment and sell it to low-income families in their district. We should be encouraging the establishment of such businesses. We should also end the practice of disposing of something, just because the colour has gone out of fashion, and as consumers we should insist on better quality, longer-lasting products.

Furniture
There are endless ways of turning so-called junk furniture into perfectly acceptable items of furniture. Cutting the legs off a kitchen table to make it into a coffee table, using a wardrobe door as bedhead, painting and stencilling a decrepit chest of drawers, are just a few examples of the limitless. The art of converting junk to style is to ask the question, if it won't do as it stands, what will it do for? Sometimes it will need just a coat of paint, other times a complete overhaul. If you are a novice and would like help, you could attend evening classes in woodwork or upholstery to guide you in the right direction.

If you want to dispose of a piece of furniture, don't throw it away. Ask around; you will be amazed at the take-up on offers of furniture. There is always a charity or an individual who will be able to make use of your cast-offs.

Garden Furniture
Use your imagination. Basic benches and seats can be made from old timbers. Old church pews make terrific garden seats, barrels make attractive plant holders, as do wheelbarrows and chimney stacks. In many ways you are even less restricted in the garden than you are in the home. Let your imagination run wild.

Jars
Why buy expensive storage jars and containers, when the humble jam or coffee jar will do the job just as well. Make jams and chutneys or pickle your own onions, then use jars for storing them. Use them to store pulses and dried fruit. Use them for nails and screws in the workshop. Sprout alfalfa and cress in them. There are a million and one uses for jars around the house, and they should only be bottle banked as a last resort.

Jumble Sales
It is interesting that many boot-sale stall holders visit jumble sales to buy things for their stalls. This is because jumble sales are an excellent place to pick up a second-hand bar-

gain. It is not everyone's idea of fun to fight amongst a sky-high pile of clothes and bric-a-brac on a Saturday morning, but it is certainly worth a rummage in terms of bargains.

Junk Food
Most fast food restaurants contribute huge amounts of paper and plastic packaging to the rubbish problem. Trials are being undertaken in fast food chains in the United States to introduce recyclable containers, but in the meantime the problem remains. Write to the restaurants who are guilty, let them know how unhappy you are at this practice, and ask them what their future plans are.

Junk Mail
The problem of junk mail is one that may be easier to deal with than you imagine. If you write to the Mailing Preference Service, Freepost 22, London W1E 7EZ, they will ensure that their member trade associations only send you direct mail on the sorts of goods you are interested in. The only problem is 'direct mail' means mail addressed directly to you. If it is addressed to 'The Occupier', they will not be able to intervene. You can choose to be excluded from all direct mail categories or just selected ones. You do not have to pay for this service.

Leaves
Don't dump the autumn leaves that fall off the trees. Keep them in a separate container from your compost for 12 months and then use as a surface mulch. They represent stored sunshine and nourishment.

Litter
Everyone is aware of the problem of litter. You cannot fail to notice the discarded drink cans, old plastic carrier bags, and fast food containers strewn around our towns and countryside. As a short-term measure, we can organize litter picks and do our bit as responsible citizens. However, we really do need to look at the reason why we generate so much litter in the first place. It comes back to our throw-away mentality. Any pensioner will tell you that litter was never a problem forty years ago; but there wasn't gross overpackaging forty years ago either. In Suffolk County, Long Island, they are

banning the use of non-biodegradable packaging at retail establishments. In Canada and in some American states you rarely see discarded cans, because they are sold with a deposit. For further information contact the Tidy Britain Group.

Metal
Much metal used in the home can be recycled and even reused. One enterprising young artist is making his living by creating sculptures from old drinks cans! Aluminium is presently the easiest metal to recycle at a domestic level. Charity and community groups are keen to organize collections when they realize that each can is worth about 1p to their funds. It is also beneficial environmentally to recycle aluminium, because it reduces energy costs to produce it by 95 per cent. Bauxite extraction (from which aluminium is obtained) is also very damaging environmentally. Information about aluminium collections (which includes cans, foil, window casings and saucepans) is available from the Aluminium Recycling Association. They will tell you how to organize, and where your nearest aluminium recycling point is.

British Steel are currently promoting the recycling of steel cans. The can-makers site 'Save-a-Can' skips around the country, where you can dispose of all your household tins and cans. These have the advantage of taking both aluminium and steel cans – they are separated by the Association when the skip is full. For further details, contact British Steel and Tin-Plate Association.

Newspapers
Everyone in the United Kingdom consumes two trees-worth of paper every year. Recycling paper makes environmental and economic sense. Yet, at present, we only recycle 27 per cent, so we have a long way to go. Newspapers are the easiest material to collect from households in bulk, but fluctuations in price have made the business of newsprint a risky one. Think of ways of reusing your newspapers before you send them to the local paper skip. Children can use them for play, to paint on, and to make paper models or papier mache (see page 143). Schools are always eternally grateful for a supply of newspapers for painting and glueing activities.

You can use it to wrap items up for storage, or make paper patterns for dressmaking. It serves well as an insulating material in your loft, a draught excluding material and even as an insulating layer between the blankets on your bed!

Oil
It is illegal to dump oil down drains or into any water course, because it causes pollution, reducing the efficiency of bacteria in the sewage system and destroying wildlife habitats. Many garages and civic amenity dumps are now offering oil recycling facilities. They are not always well publicized, but ask around and you will find a local facility; then you won't be contributing to the 50,000 tonnes of oil unaccounted for each year.

Plastics
Plastic packaging alone makes up 20 per cent of domestic waste in this country. The use of mixed plastics makes recycling extremely difficult. A very small number of retailers offer a refill service (such as the Body Shop), but the take-up is small. Most plastics are not biodegradable and cause pollution problems in many areas. As consumers we can do something to help prevent this major problem. We can refuse plastic carrier bags in shops (take a shopping bag or basket). Try to avoid goods that are overpackaged, and let the retailers and the manufacturers know how you feel about this wasteful practice by writing to them.

Playgroups and Schools
These are both marvellous receivers of unwanted packaging (yoghurt cartons, egg boxes, and so on, to make models from), clothes (for the dressing up box) and books and toys. Contact the Pre-School Playgroups Association for details of your local playgroups, and contact your local schools.

Scrap Stores
These are also known as Resource Centres, and they are a wonderful example of how a successful recycling project can be established in a community. They collect and store a wide variety of waste materials donated by factories, large firms, small businesses and public bodies, and pass them on to groups in the community concerned with children. Children

benefit from the many play materials available (which would be expensive commercially). Local companies can cut their waste disposal costs and improve their standing in the community. Although they are not strictly related to the subject of junk and recycling in the home, they do put a focus on using someone elses 'junk' for a positive community use. For further information contact the Federation of Childrens Resource Centres.

Skips
When you think of all the skips sitting on drives, full of discarded housing material waiting to be taken to overstretched landfill sites, you begin to realize what a waste they are. It is estimated by the Building Research Establishment that 9 per cent of bricks and 10 per cent of timbers are wasted before the house is even built! Much of the material in skips could be reused or recycled. It is not unusual for people to discreetly take junk finds from skips (although this is illegal). Think carefully before you hire a skip for any reason. Don't be tempted to use it as an excuse to dump all your so-called unwanted items because it's less hassle than recycling it carefully.

Spectacles
Don't throw your old spectacles in the bin. Many charities and opticians send used glasses to Third World Countries to be sorted and distributed to needy people there. Make enquiries at your local charity shop or opticians.

Tipping
Waste ground is all too often the target for dumping of mattresses, fridges, bikes, and any other number of household items that have come to the end of their useful life. Please inform the waste management department at your local council if you see this happening.

Toxic Waste
Toxic waste does have a place in a book about the green home, even though we may not feel we are to blame as individuals. We are responsible for industrial toxic waste, because we are the consumers who are buying the goods that create the pollutants. The most obvious thing you can

do as an individual is to reduce consumption, then you will be contributing less to the production of the stuff. Secondly, encourage others to look at the problem by drawing attention to it. Find out if tankers carrying toxic waste travel through your town, and if they do, kick up a fuss. Greenpeace Toxic Waste Division will offer advice if you encounter a specific problem, such as suspected illegal (fly) tipping. Or you could join a CAT (Communities Against Toxics) group, who are constantly campaigning against the problem.

9 Hobbies and Home Crafts

Introduction

Dissatisfaction with mass-produced goods has led to a huge revival of interest in traditional crafts, allowing everyone to experience the lasting enjoyment and satisfaction of creating beautiful objects with their own hands.

John Seymour

Traditional crafts and hobbies might be thought of as 'cranky' or 'ridiculous' by many people these days, but the art and skill of many of our crafts may well be lost forever unless we develop a healthy respect for knowing how and why things work and how to do it ourselves.

Many arts and craft fairs still exist, providing interesting days out for punters and real excitement for those who have spent time creating something of beauty and skill that they want to show off. Many people take their hobbies seriously by entering competitions for prizes (which can be very lucrative), while others prefer to sell their goods for charities or to raise money for their local associations, like the Womens' Institutes or local church groups.

Crafts can save people money on products that they would never generally have been able to afford. A carefully hand-sewn, and appliquéd bedspread can be bought for £300 and yet the cost of materials may be only valued at £30 or £40. The dedicated work that goes into making it however, may be priceless.

A good hobby that saves money is breadmaking. The rich aroma of freshly baked bread winding through a house gives it an immensely warm and inviting atmosphere and the bread itself hardly needs an invitation to be eaten. Many people who make their own bread use organic flours, which give the bread a superior taste as well as cutting down on the additives and unnecessary pesticides found in the ordinary varieties. Not only that, but the bread can be made in batches and frozen, so it only needs to be made each week or so. Imagine all the money and packaging you will save!

Positive Effects of Crafts
Crafts and hobbies can be much more than breadmaking and sewing however. Traditional skills like spinning, woodturning, pottery, weaving and candlemaking are coming into their own again, and bookshops sell literally hundreds of beautifully illustrated volumes giving ideas, recipes or step-by-step instructions. Stained windows can make an ordinary piece of glass look fantastic, and careful woodwork and painting work increase the value of your home.

Finding out how to make things and how they work is a skill in itself. So many of us buy the first available 'off-the shelf' product, all wrapped and ready to use, that we completely forget where the motivation for the product came from in the first place. The joy and hours of fun that small children get out of creating their own batik scarves or making wooden spoons should remind us that we will treasure and understand our own needs more by doing it ourselves.

Creating crafts for others in your local area helps the environment as well as your self-esteem and finances. By using local resources available cheaply you can cut down on the amount of transport needed to bring in products from outside the area; this in turn cuts down on the amount of pollution that heavy lorries are responsible for, creating a better local environment for everyone. The consequences of this cannot be underestimated. Just imagine all the energy required to bring copies of wooden sculptures or pottery bowls from overseas. Or the carbon dioxide created by bringing handprinted fabrics and handknitted jumpers from India; and what about all those imitation knick-knacks that sell so cheaply on market stalls? Often these products are manufactured on a large scale, using plastics and synthetic materials that undermine the real beauty of the design. Toxic or leaded paints may be used, and chlorinated solvents or asphyxiating lacquers often finish a piece; all of these create environmental pollution in their production and in their disposal.

Add to that the exploitation of low-paid workers, especially the grossly undervalued women and children in the Third World, and buying crafts from overseas becomes a socially and ecologically unacceptable way of consuming our dwindling resources. Some groups are making vast strides in educating the affluent consumer about Third World debt and socially conscious shopping, however. Groups like Oxfam and Traidcraft specialize in showing us another way to consume. They set up direct links with the producers, buying products mainly from small cooperatives where profit is shared and where they produce goods that are useful to the buyer, provide a decent wage to the producer and do as little damage to the environment as possible.

Cultivating and learning local crafts can help the environment in other ways too. Bee-keeping could prevent the extinction of particular varieties of bees, and growing and cultivating apple trees can have an enormous effect on the apple, once the beauty of British gardens. (Several hundred species are now facing certain extinction; British apple growers used to boast over 6,000 varieties of apple but only nine or ten are grown commercially and are available in most shops and supermarkets.)

Cross-pollination and fertilization keeps the birds and the bees busy in the spring, and the fruit trees provide a nutritious food source for many animals in the autumn. An abundance of any sort of fruit or vegetable in your garden means that you can make jams, jellies and chutneys.

A Pastime – and a Community Resource

In today's fast-moving world many people will reject the idea of pursuing any type of craft or hobby. They may prefer to shop for their goods instead of making them, or they may have too many other responsibilities, like children and work, to allow them to do little more than the minimum they need to survive. Their time may be as important to them as their money and energy and there is nothing that anyone can say or do that will make them change their minds. But some people do have the time and want to learn new skills and crafts. Our society should encourage and support these people as much as possible; they should also be encouraged to share their skills and teach others, passing down valuable skills to the next generation.

One group in western Canada used the idea of sharing crafts and skills to their benefit. They set up a community databank of various people's skills. Each time a member wanted or needed something, something mended or an unusual gift for a friend, they would call up the databank and see who could help them. No money exchanged hands, instead the recipients would offer something they could do for free in return. It might be dressmaking, or help with home decorating or accounting. They would work for roughly for the time needed to pay back the cost of the original need. This way the community benefited greatly from people who were willing to share their hobbies and skills as needed. The project was so successful, however, that the local authorities moved in to close it down as they were afraid of the effect that such sharing would have on the Canadian economy!

Using natural materials to make the things we need ourselves is a great pleasure in itself. Stone, wood, leather, natural fibres like hemp, flax and cotton, clay, rocks and gemstones, crystals, metals and iron can all be made into beautiful artefacts and objects. The discipline involved in creating from natural materials means that we can create unique and varying objects and are no longer tied to industrialized production methods using plastics and synthetic chemicals. The pleasure felt by the craftsmen and craftswomen in producing fine goods of value for us to buy is matched by the pleasure that the buyer gets from the product itself. Handcrafted work is more valuable; it does cost more, but well-made, quality work will last far longer.

Our dream of progress was that hard manual work would be replaced by the wonders of automation and technology. Fifty or one hundred years ago we believed that increased mechanization would be the answer to the poor workers' nightmares. But the progress dream was shattered when men and women learned that they would have to work just as hard in a factory. Now we develop hobbies after working hours to satisfy our need to be creative.

Hints for Getting the Most out of your Hobbies

1. Choose crafts that excite and interest you; do it for

yourself first and for the pleasure of others only if you like it.

2. Do not be over-ambitious to start with. If you take on mammoth tasks that you will soon get disillusioned or bored with; try small things first and see if you enjoy them.

3. Use locally-produced resources.

4. Use only non-toxic components, reject plastics and pvc.

5. Try varnishes and paints that are organically produced, and do not use excessive chemicals.

6. Use homemade products as often as possible, from jams to shampoos.

7. Buy and support locally-produced crafts.

8. If you buy from overseas, buy goods from companies and organizations that support sustainable and equitable conditions for workers.

Directory

Appliqué
Appliqué is an artistic way of recycling, or at least reusing leftover materials. It involves the sewing of small cut out pieces of fabric onto a larger background piece. It is often used in conjunction with quilting and patchwork. Appliquéd cushions add an elegant touch to any living room, and recycle scraps which may otherwise have been thrown away.

Aquariums
The practice of keeping and breeding tropical fish is increasing in popularity in this country. The problem is that these aquariums deny exotic tropical fish their natural habitat. Up to 60 per cent of imported tropical fish die within a year of purchase. A boxed tank with artificial rocks and plants will never replace their natural home.

Basketry

Until quite recently there were many thousands of basket-makers in this country, but the numbers have now dwindled to a few hundred. Baskets can be used for a multitude of purposes, including storing, carrying and displaying. Basket-making requires little equipment and is easily learned. Cane, willow or rush are the most common materials used, and it is quite possible and most fulfilling to gather them yourself. Willow is the most popular choice of material in this country, and, reassuringly, it is a sustainable crop, yielding an annual harvest which grows well on marshy land unsuitable for other crops. For further information, contact the Basketmakers Association.

Batik

Batik is an ancient Indonesian craft dating back 2,000 years. It is achieved by applying a coloured design onto textiles and waxing the parts that are not to be dyed. Batik can be used for all sorts of things, from wall hangings to t-shirts to bed covers. Every design will be totally original, and you will have created something unique.

Bee-Keeping

Nothing tastes as delicious as home-produced honey. There are 250 varieties of bees in this country alone, but because of pesticides and wet summers, the bee is now under threat. It is important that we cultivate the art of bee-keeping, to ensure that we prevent the extinction of certain varieties. For further details on bee-keeping, contact the British Bee-Keepers Association.

Beer-Making

The best home brew without a doubt is made from real hops (which will grow on most land, provided it is well dunged and is not waterlogged) together with malt, yeast, sugar and water. Commercially produced kits contain cocktails of chemicals and are usually wastefully overpackaged. CAMRA (The Campaign for Real Ale) have a book called *Home Brew*, which advises the best way to produce a natural home brew. Contact them for details.

Bread-Making

Once you have mastered the art of bread-making, you will

never want shop-bought bread again. The trick is to use fresh, wholesome ingredients. Hunt out a supplier of fresh organic flour. Use fresh yeast, and locally-produced honey to help it ferment. The experience of kneading bread by hand can be very therapeutic, but if time is at a premium, and you have a food processor, do use it. The results will be equally delicious and your house will smell of freshly-baked bread, arguably one of the most tantalizing aromas in the world.

Calligraphy
Calligraphy is the art of handwriting. In these days of typed or printed language, the art of handwriting is often neglected. By using calligraphy you can give letters and cards a really personal and individual look, and help to keep a very old tradition alive. Any good stationer will stock a selection of pens and nibs suitable for use, or be really traditional and use a quill and ink!

Candles
Although they are no longer a necessary form of lighting, candles create an almost mystical environment. They provide a focus for meditation, and an atmospheric alternative to the static glare of the electric bulb. Making candles at home for fun or profit is surprisingly easy and inexpensive. Don't be conned into buying one of the expensive kits that are available – you can make candles successfully from equipment gathered together at home. The Association of Candlemakers will provide you with information and a list of suppliers of equipment.

Collage
Collage uses scrap materials – paper, pasta, beads, leaves, twigs, in fact anything at all – to create pictures, wall hangings and so on. The appeal of collage is that you can create something truly original without special training or talent. The only specific equipment that you will need is a supply of glue. Try making your own glue from cornflour and boiling water, or if you are using a commercially-produced glue, make sure it is water-based, which is less harmful. If you have problems, you may find that your children are well versed in the art of collage and will be more than willing to lend a hand!

Dress-Making

Making your own clothes is worth serious consideration, even if you have never done it before. Second-hand reconditioned sewing machines from major retailers are usually good bargains, or you could borrow one from friends. There are many good patterns around, or you could even design your own. By shopping wisely for fabrics (try to buy natural materials such as cotton or wool), you can save money and help the environment. It also gives you the opportunity to revamp old clothes, either from your own wardrobe or from jumble sales and charity shops, at a cost of next to nothing.

Dyes

The ancient craft of using natural dyes is being revived as weavers and knitters rediscover the softer and subtler shades that only natural dyes can offer. Birch, lichen, juniper berries, saffron and spinach are just some of the sources of natural dyes. The dyes are extracted either by soaking the bark of leaves in water, or crushing berries, flowers, and so on. The dye is then placed in a bath (an old baby bath will suffice) and a cup of sea salt dissolved in boiling water is added. This is heated gently and the yarn immersed in it. It is left to simmer for about 40 minutes, by which time the yarn should have taken up the colour.

Embroidery

Embroidery is an ancient and skilled craft that can be used to give a personal touch to lots of everyday items, such as handkerchiefs, tablecloths and pillow cases. The only equipment you will need is a supply of needles and embroidery threads. Smocking is another form of embroidery, one that provides a solution to the problem of how to make a garment fit snugly without restricting movement. For further information contact the Embroiderers Guild.

Flower-Arranging

The mass production of flowers for the cut-flower market has resulted in the use of dangerous pesticides and fungicides to meet ever-increasing demands. Flowers are also now bred for size and appearance, at the expense of natural scent and colour, leaving us with a cloned substitute for what was once a wonderful wild bloom. Picking wild flowers is definitely *not* the answer either. Leave wild flowers in their environment as part of the natural ecosystem. If you want to do flower arranging, use flowers from your own garden, or dried flowers and grasses, which will last for anything up to three years and will look equally attractive.

Glass Engraving
Glass engraving is both an art and a craft, an exciting medium for the artist and a hobby for the practically minded person. Engraved glass is not only beautiful, but highly individual, and the most memorable pieces are crafted by hand using diamond point equipment. Hand engraved work always has a slight sheen or sparkle due to the multitudes of tiny cuts made by the diamond in the surface of the glass. This is a reminder that the glass was engraved painstakingly by human hand and makes it all the more precious. For further information about glass engraving, contact The Glass Manufacturers Federation.

Jam-Making
The biggest problem with buying manufactured jams, jellies and marmalades is that you are buying a product which is made from fruit that may be sprayed with chemicals, and that is high in sugar and additives. Making your own preserves means that you can use organically-grown fruits, reduce the amount of sugar used, and help to keep a tradition alive. Use half the sugar recommended in the recipe, the jam may seem a little runnier, but you'll save money and be healthier.

Jewellery
Natural objects have been used to adorn the body since time began. Nuts, seeds, twigs, shells and feathers can be used to make attractive and original jewellery. Making your own jewellery is also an opportunity to recycle materials like newspaper (by making papier mache jewellery), fabric scraps, buttons, beads and corks. Most haberdashery departments and craft shops will sell you clasps, studs and other equipment.

Kite-Making
Kite-flying has been a national pastime for Koreans, Chinese, Japanese and other eastern cultures for thousands

of years. Today the art of making and flying kites is still thought of as predominantly an eastern pastime. However, those in the West who have discovered the thrill of kites are enraptured by it. Making kites is no less an art than any other and is certainly not thought of as 'toy'-making by the converted. Kites can be made in many different shapes and sizes, but for the novice, the traditional diamond-shaped kite would probably be the most suitable choice. For further reading try David Pelham's *Kites* (Penguin).

Knitting

Knitting need not be just a hobby, it can be a form of self-expression. You can create your own unique designs from materials that have been derived in a way that is environmentally sustainable. Avoid acrylic/wool mixes that are derived from petrochemical derivatives, and go for pure wools and cotton. Old jumpers picked up at jumble sales can often be unpicked and re-used. Initial outlay will involve needles and patterns, but, looked after well, both will last you a long time. Books and magazines with pattern ideas and step-by-step instructions are available in abundance.

Lace Making

It looks fiddly and time consuming, and it is a difficult skill to learn, but once mastered it can become a consuming passion. There is something magical about producing intricate patterns from lace, and you will find that your skill is in great demand. The two main ways of making lace are needlepoint and bobbin. Most people use a combination of the two. Although you can teach yourself lace making from a book, it would be advisable to seek out lessons and learn from an expert. Contact The Lace Makers Federation for further details.

Leatherwork

Working with leather requires a high degree of skill, and traditional specialist tools and materials are required to achieve good results. It is therefore an expensive hobby to embark on. From an ethical viewpoint, it is also worth considering the use of leather as a material. Vegans and many vegetarians believe that using leather is exploiting animals, and so look to hardwearing cloth such as canvas as an alternative to leather. For further information contact The Vegan Society.

Marbling
When you open an old book you are often struck by the beautiful marbled effect on the dust covers. Paper marbling dates back to the crusades, and has been used by bookbinders since the seventeenth century. It is an easy craft to learn. Oil or watercolour paints are floated on water (which has been thickened with wallpaper paste) and a sheet of paper is laid upon the floating colours to catch them. When the sheet is pulled away, the water is clean and all the colour is on the paper. Practise marbling and use it for creating individual wrapping paper, cards, book-sheets or anything else you can think of.

Murals
Wall murals are a cheering sight; they add life and colour to often drab, grey surroundings. They also offer an opportunity for collective creativity, often involving the local community as well as artists. Get your local council to support the idea of a community mural project. You could paint a pictorial history of your local area, or depict a famous local event. It would be a good chance to get to know other local people.

Origami
Anything that makes us reassess the value of a much-used material has to be a good thing; it can prevent us from taking a precious resource for granted. Origami, the Japanese art of paperfolding, does just this with paper by turning it into an artform. It can be used to make presents such as models, or mobiles, and masks for carnivals and parties. *Origami* by Zulul Aytine Scheele (Hamlyn) is a good starting point for beginners.

Paper Making
Recycling your own paper to create original handmade paper for personal use is relatively easy to learn, if slightly messy! Paper-making kits are available, or you can have fun gathering together your own. Write to Friends of the Earth for their information sheet on paper making.

Papier Mache
Papier mache is a highly versatile medium, using cheap and readily available materials. It involves pulping paper (pre-

ferrably newsprint) by mixing it with water and a thickening agent (flour or wallpaper paste), draining it, and then modelling it around a mould, which can be plasticine, a balloon, or any number of other objects. The results can be as simple or as sophisticated as you wish, from bowls to elaborate toys and sculptures.

Patchwork
The saying 'necessity is the mother of invention' can be applied aptly to the craft of patchwork. It was a skill developed by New World settlers who used every last bit of material that they took with them to great effect because they had to. Nowadays we tend to buy materials to give a coordinated effect, but wonderful patchwork can be achieved using genuine scraps of material to create something out of waste.

Pottery
The versatility of clay means that there are countless different ways of making (and decorating) pots. Although mechanization has taken over from hand-work to a great extent, there is still a good market for hand-thrown pottery. As people become disillusioned with mass-produced products and learn to value the skill of craftsmen and women there will be a return to hand-crafted goods. If you wish to learn pottery, enrol at your local college, or seek out a local potter, who may well be willing to teach you.

Printing
Block printing has been a way of decorating paper and fabrics for many centuries. Traditionally, wood was used, but you can use potatoes, corks, linoleum cuts or anything else that you feel would imprint well. The aim is to obtain an impression on paper or fabric from a block coated with paint, dye or ink. It can be used simply, to embellish stationery with a single motif, or daringly to design patterns for clothing. It is a fun craft to explore, and one that you can share very successfully with your children.

Quilting
Quilting is defined as the stitching together of two pieces of cloth with soft padding material between them. Quilted items such as cushions and bedcovers can add style and

individuality to the green home. There are twelve different types of quilting, and it is quite difficult to learn. Contact your local Women's Institute, who will probably be able to advise you on classes in your area.

Rug Making
Growing concern about conservation and the recycling of materials means that we may well see a revival of the rag rug. Until the advent of the fitted carpet, rag rugs were a common sight in most homes. Often the whole family would be involved in rug making, so it became a focus of family life. Rag rugs are made by hooking pieces of fabric through hessian. By using colour and texture of different fabrics, you can create an attractive piece of furnishing.

Spinning
The hand spinning of wool and fibres is one of the oldest and most basic of crafts. In her book *The Craft of Hand Spinning* Eileen Chadwick includes information about the sources and quality of different sorts of yarn. You can either use a hand spindle, which you can make yourself (from a round piece of dowelling and a rubber heel) or a spinning wheel, which will cost you anything between £150 and £3,000. Whether you use a spindle or a wheel, the wool is first prepared by 'carding' to create a roll of fibres to spin. The process can be very relaxing. Try a home-made spindle to give you a taste of spinning, and look in one of the many books available for a list of suppliers.

Sponging
Using sponges to apply paint to our walls and furniture has become very popular in recent years. However, before you reach for the sponge to decorate your lounge or your chest of drawers, you should consider that the removal of sponges from the Mediterranean is stripping the sea of natural life, and the sponge is threatened with extinction. Use rags and try rag-rolling as an alternative instead. The effect will be equally pleasing with much less damage to the environment. Don't forget to use lead-free non-toxic paint! (See our DIY chapter).

Stencilling
Stencilling is any method of decorating that involves apply-

ing colour through shapes cut out of a sheet of impervious material. Anyone can do it and you can make your own designs (you can buy ready-made ones, but its much more fun to make your own). The only tools you will need are card, brushes and paint. Start with stencilling something simple, such as a tray, and you could find yourself stencilling every wall in the house.

Toy Making
Millions of pounds are spent on advertising every year to try to persuade children to want, and parents to buy, toys that are expensive, badly made, and often contain toxic dyes and colourings. The whole question of targeting advertising at children of a very young age is open to debate. In the meantime the obvious solution would be to make your own toys. Toy making incorporates a number of crafts; needlecraft, embroidery, papier mache and woodwork to name but a few. You will be able to choose your own materials from sustainable sources, recycle materials, and gain a great deal of satisfaction knowing that you are giving your child a safe toy made lovingly by yourself. Of course, encouraging your children to take up craft making is another option. They will enjoy the creative challenge of making things as much as you.

Weaving
Weaving is one of the most ancient and fundamental of crafts. You will require a loom, which can vary from a simple home-made model, to a large floor loom used by the professionals, and a shuttle. This is the basic equipment. Almost any yarn can be woven, but try to stick to natural fibres like cotton, linen and wool.

Wood Work
There is something about wood that makes the woodturner or woodcarver into the poet of the craft world, waxing lyrical about the quality of the wood, the grain, the smell and the texture. Woodwork appears to make us more aware of how beautiful and important trees really are. Each species has its own special characteristics, and these are in evidence in wooden items. We should avoid using tropical hardwoods, even for small craft projects, because of the destruc-

tion of the rainforests. We should look towards using native hardwoods forested sustainably. For more information read Friends of the Earth's *Good Wood Guide*.

Writing
When Edith Holden wrote her nature diary in the late nineteenth century, little did she imagine that it would become a bestseller in the form of *The Country Diary of a Victorian Lady*. She did not set out to impress a reading public – her diaries were simply her thoughts on her natural surroundings, a subject very dear to her. Sadly, many of the plant and animal species she refers to are becoming increasingly rare. Which is why more of us should consider keeping a similar diary about the wildlife, flora and fauna that we encounter (yes, even as a city-dweller). Recording changes in nature will make us much more aware of our natural surroundings. And who knows, in a hundred years from now, someone may be looking at your diary and imagining how life used to be!

10 Furniture

Introduction

Have nothing in your home which you do not know to be beautiful or useful.
William Morris

'I'm just part of the furniture' is a phrase often used when we are feeling ignored. Furniture in our home is often taken for granted; we are surprised when we go into a home which has little or none save a few cushions. Sitting on the floor is thought by some to be 'bohemian', yet for our backs it is the best way to sit. Very few native peoples sit on chairs; most sit or crouch on the ground, providing a better position for the back. Sofas, likewise, are not particularly good for the spine.

Bad Design

Most furniture, in terms of human health, is badly designed. It is built to suit machines, not people. It is also badly built; the materials that are used to make it are the cheapest possible, and the quality of construction is generally low. In the manufacture of furniture, no consideration is given to the workers who make it or the environmental effects its construction could have.

Each individual component in a typical piece of furniture is full of environmental problems. The frames are often constructed from chipboard, which is full of glues made from chemical solvents. The foam used for most padding before 1989 was made by a process which emits large amounts of CFCs (this foam is still being produced in many places outside the United Kingdom). The coverings are frequently made from synthetic fibres which are made from petroleum-based products, and the fabric is then finished with fire retardants and other chemicals. All these form a cocktail of chemicals which give off toxic gases in the home.

The most common constituent of all the chemicals is formaldehyde, which is an irritant to the eyes, nose, throat and lungs. It can cause skin reactions. Exposure to formaldehyde has been shown to cause headaches, depression, dizziness and loss of sleep. It aggravates minor illness such as a cold, and can trigger more major illness. A recent study on the toxic effects of formaldehyde concluded that one in five people could be sensitive to it. Most people can smell formaldehyde at concentrations of 0.1 parts per million, yet some homes register formaldehyde at 1.0 part per million. Formaldehyde pollution comes from the materials used in the home and its construction. It is used in large quantities in foam insulation, chipboard and plywoods and poses a serious health risk because it remains active for months or even years.

The glues used in furniture construction are a problem, as they often contain solvents and formaldehyde. These substances emit gases for extended periods of time. Epoxy glue is particularly hazardous and should be avoided. The hardeners are oxidants, and in the hardening process the chemicals react together to form another compound. As a by-product of this reaction a gas is given off. This process continues after the glue has set and can start again if the glue is heated. Sometimes a warm house interior is all that is needed to restart the process.

Wood

Wood can be divided broadly into two categories: hardwoods and softwoods. This is a botanical division, referring to their respective types of propagation; these terms do not classify how the wood can be used. The forests of the whole world are rapidly disappearing; England was once covered with oak trees. Many of these went to make the ships for the navy and much of the furniture we now see in stately homes. Many poor countries are selling their timber to help pay their national debts. When the forest is cut down the animals have to find new homes, and the undergrowth dies as it no longer has the protection of the forest canopy. Destruction of the forest in this way does not allow the forest to regenerate, as too much is cut down too quickly. Reforestation in such regions can only occur if trees are deliberately replanted: propagation cannot take place where vast areas are suddenly stripped of the parent trees. There is a tendency to replant with fast-growing trees rather than native broadleaf, as the faster the trees grow the sooner investors can see a return for their money. The forests we are cutting down are disappearing seven times faster than we are replanting them.

Trees help to keep the earth's water balance stable by holding water in their roots; the roots also help prevent the soil erosion, which is affecting ecosystems all over the world. At present, besides destroying the forests, we are also making it more difficult for the remaining trees to survive. Many forests in the Northern hemisphere are being destroyed by acid rain: in Germany at least 50 per cent of the trees are affected, and in England, 67 per cent of the conifers show signs of damage.

We can plant sustainable forests with a tree being planted for every one cut down, but planted forest cannot reproduce the complicated and fragile ecosystems of a forest which developed naturally. Rainforests are the richest source of life on earth. Over 50 per cent of the earth's species live in these forests, yet we are destroying them at the rate of 100 acres per minute. In the last ten years we have cut down 25 per cent of these forests. Many, many acres of the forests are being burned rather than cut. This releases tremendous amounts of carbon dioxide, which is the heaviest contributor to the greenhouse effect, also known as global warming.

Ironically, the very things which were designed by nature to cope with carbon dioxide, the trees themselves, are the most immediate victims of the destruction. We may think we are far removed from this environmental crisis, but in fact we are also very directly affected by it. The rapid destruction of such vast numbers of the world's trees is causing the earth's temperatures to rise; trees are the planet's foremost natural cooling agent. Ecosystems all over the world are changing, the polar icecaps are slowly melting, and unless it is stopped life on earth will change forever – and we will be the losers. We have already lost hundreds of species of plants and animals in the rainforest decimation; nature not only enriches our world with its beauty and complexity, it also provides us with unique sources of knowledge. The tropical forests contain plants which exist nowhere else and which have provided researchers with new medications to fight disease. The keys to developing many more such drugs almost certainly lie in the rainforest, but we are running out of time.

There are, however, things that *you* can do to help save the tropical forests: do not buy tropical woods; avoid buying leather products from areas where rainforests are being destroyed (many of the forests are being sacrificed to raise cattle that the world does not need); cut down or eliminate your meat intake; try to support Third World industries that depend on the forests (such as nut harvesting); help to support the local people in other ways (buy their crafts and products). Last, but not least, spread the word!

The building and furniture industries are major utilizers of these endangered timbers. This is why it is so important to make the most careful and well-informed decisions about your own use of wood, and to make sure that the wood products you purchase are well cared for and long lasting. Wood is an extremely popular material for furniture because of its sturdiness, durability, and the ease with which it can be worked. It also has tremendous aesthetic appeal: it is pleasant to the touch, the grain and warm rich colours are visually attractive, many woods have a fresh, fragrant aroma, and many people find it more atmospherically 'friendly' than metals, stone or synthetics.

Not so long ago, our forests supported chair bodgers, who worked in beech woods making windsor chairs on the spot. They used to work in pairs, one chopping the pieces, the other turning. This is a craft, along with cane weaving, which has nearly died out in this country. Careful management of our forests could help regenerate such traditional arts and creative employment once again.

The best sort of furniture to use is that which has been made using the old technique whereby the furniture is held together by special joints rather than glue. It should be made of solid wood, and waxed rather than varnished, so that the wood can breathe.

Points to Look for When Buying Furniture

1. What is the source of the timber?
2. Was the forest sustainably managed?
3. How are the pieces held together? Does it have natural wood joints and little glue, or does it have a lot of glue, particularly epoxy?
4. What is the filling? Is it made of natural or synthetic fibres?
5. What is the covering? It should be made of an unbleached fabric; you can place a decorative patterned loose cover on top.
6. When buying bamboo furniture, avoid those pieces with dark flecks, as this means they have been treated with harmful chemicals.
7. Try to obtain furniture that has not been varnished.
8. For those who have problems with allergies, an alternative mattress can be bought made of latex.
9. When buying furniture try it out; sit on it for a long time to make sure it is really comfortable.
10. Are the items of furniture made out of endangered rainforest woods such as teak, ebony or mahogany? If so, shop around and seek alternatives: there are many other beautiful woods available.

Some Information about Different Woods

Hardwoods
Alder This is a native wood that is very good for making brooms and brushes as well as toys. It is light, soft and easy to turn. It does not have a very long life if used outside.
Apple Only use applewood if the tree has come to the

end of its life and another has been planted. It turns well, and so is good for making household utensils.

Ash This is an exceptionally strong wood which is nevertheless very flexible, and thus is highly valued for furniture making. It also stains very well.

Beech Another very strong European wood which bends well. It is the United Kingdom's main hardwood. Beech is used to make chairs, desks, doors and block flooring.

Elm Dutch Elm is 40 per cent tougher than English, but supplies have been greatly reduced by disease. It is used in chairs, settee frames and for cabinets. It has good bending properties and can also be used in flooring.

Maple Some maple has an attractive, swirling grain called bird's eye. It is very strong and very good for furniture making. It is also useful for worktops and panelling.

Oak One of the most commonly used woods for furniture and stairs. It is also used in the production of window sills, doors and fencing. It is a good alternative to teak and mahogany (i.e., woods from threatened tropical forest) for dining room furniture.

Walnut Most often used in the veneer of high quality tables, especially in the boardroom.

Soft woods

Cedar Used in construction work, often for roofs. It is very strong and resistant to weather. Its aromatic qualities make it an effective insect repellent, and so it is often used to build closets, drawers and cupboards.

Pine Used throughout the house both outside as windows and doors and inside, as furniture and for flooring. It needs to be sealed before being treated with a natural wax.

Spruce Used as a construction timber for roof beams, floors, planking and joinery. Very resistant to weather and fungus.

Yew A strong wood used in furniture, especially in bedrooms.

Woods that Should be Avoided as they are Endangered

Ebony, **teak**, **mahogany**, and **rosewood** are some of the best known rare woods in this country but all are threatened.

If you need to use timber think about its ultimate purpose. Tropical woods are indeed beautiful, but their beauty is only one of the reasons we should be preserving rather than cutting them.

Other woods *can* be used instead. Whatever wood you do end up choosing, make use of it in such a way that it will last for at least a hundred years – enough time for another tree to at least start growing.

Remember that trees require many years to mature; they do not magically renew themselves as we deplete their numbers.

Directory

Bamboo
This is a good alternative to hardwood furniture. Practically anything can be made from bamboo. Most bamboo comes from India and China. It is a fast-growing crop – the world's fastest – capable of growing 1 metre a day. Pesticides are not used very much in cultivation, as infestations can be controlled by cutting. The problems come once the crop has been cut, since bamboo is a major factor in combatting soil erosion; and it is frequently treated with dangerous substances before it has been made into furniture. DDT, and Rangoon oil, a petroleum-based product are both used as preservatives. Rattan is another natural product which can be used for furniture making, as can some forms of willow. These are cheaper alternatives to solid wood furniture and just as pleasing to the eye.

Baths
The best quality baths are made from enamelled steel or enamelled cast iron; both materials are acceptable.

Furniture

Fibreglass baths are made with polyester resins and various hardeners and additives which give off gas and contaminate both the air and the water. Fibreglass is also less durable than steel. The European bath is designed to allow the bather to stretch luxuriously to her or his full length. This may feel wonderfully relaxing and decadent, but unfortunately it is a most inefficient use of resources. The large surface area created by this type of bath promotes the rapid cooling of the bathwater. A bath designed in Japan but now available in Brtain is much more energy efficient. It is a sit-down tub with high sides, in which the water comes up to the bather's shoulders. The far smaller surface area helps conserve heat, thus reducing fuel consumption – and your hot water bill!

Beds

The ideal bed frame is made from solid wood. The wood used should be from a sustainable source and properly dried. It then needs to be finished with a natural varnish, like shellac. This both protects the wood and makes it undesirable to beetles. The mattresses should be made of natural fabrics and materials. There is some evidence to suggest that mattresses made with metal springs cause geomagnetic field distortions. This means that as we sleep our nerves have to constantly re-adjust to the magnetic impulses coming from the springs. For the more sensitive this can cause restless sleep.

Beeswax

This produces a soft, sweet-smelling sheen on wood. It protects all wood surfaces, old and new alike. Commercial beeswax polishes mix the beeswax with turpentine which is an oil derived from balsa wood. This mixture is very good for floors (an added benefit is that beeswax is antistatic). Beeswax polish can be made with 1 oz beeswax and $1/4$ pint of turpentine. Scrape the wax into shavings and leave it to dissolve in the turpentine. This is a slow process and may take a few days. When the wax has dissolved, give the bottle a shake and use in the normal way. As good wood furniture is expensive, it is worth looking after well. Wood needs to breathe; therefore, it should not be sealed, as the majority of wood seals are based on polyurethane, a plastic which does not breathe and does not biodegrade.

Composite Boards

Fibre boards
Made from dried wood and vegetable fibres without the use of adhesives. They come in different varieties of thickness but are usually treated with formaldehyde.

Plywood
Is made from several layers of veneer bonded together with glues. The glues often contain phenol and formaldehyde.

Blockboard
This is made of blocks of wood sandwiched between two thin sheets of wood. Again, glues and chemical solvents are used in their production.

Chipboard and Medium Density Fibreboard (MDF)
Most European furniture is made from wood chipboard or MDF. Chipboard is made from wood chips blended with synthetic resins and pressed into sheets. It contains urea formaldehyde and resins which can release contaminates in the home. With this sort of product it is very difficult to isolate the individual constituents, so be very careful when choosing. Chipboard is manufactured according to the requirements of its application. Some people consider it a viable alternative as it is made with softwoods from sustainably grown sources.

Doors

The door is the object which literally welcomes you to the house. One choice for the environmentally correct door that also looks good and is sustainably grown is the native British oak. Unlike the 1.5 million tropical hardwood doors imported into Britian every year, its manufacture has not contributed to the tragic destruction of the earth's rainforests. Most modern doors are made from chemically treated wood and finished with lacquers containing volatile and toxic solvents. Wood preservation treatments use long-acting poisons which can cause illness and even death. The natural oak door is made from mature oak which has been limed to emphasize the grain and finished with natural oils and waxes, giving the wood a mellow glow.

Fillings

Choose from untreated cotton, wool, jute, kapok and coir. Some animal products are used for filling, such as horse hair and feathers; these are unacceptable to some people on moral grounds, to others because they can cause allergies. Polyurethane foam, as of February 1989, was banned in the United Kingdom as a filling for furniture. Even though you cannot now buy new furniture containing this material, much second-hand furniture and many items already in your home may contain it, and therefore there are some things about it that you should know. It is a serious fire hazard because it is extremely flammable; when it burns, it gives off noxious gases which can kill. Polyurethane foams also give off gas when they are new, causing a build up of static and formaldehyde gases in the home. In their manufacture they produce CFCs and other gases which are released into the atmosphere.

Frames

Metal
When choosing metal frames, care should be taken with the paint used; metal furniture frames are often sprayed with paints containing toxins. Metal products are very energy intensive in their construction. They also cause considerable local pollution in the districts where they are mined. If possible they should be avoided as they are not biodegradable.

Hardwood
These contain small quantities of glue but are otherwise acceptable.

Softwood
These are often sealed with paint varnish or lacquer. Varnishes are very toxic, as are some paints. Try to buy furniture that is finished with natural wax or varnish.

Plywood
These contain formaldehyde, and careful attention should be paid to the sealing. Shellac is a very good natural sealer.

Chipboard
These are the worst offenders, as they contain the most glue.

Furniture is an aspect of the household where we have considerable of control over what we buy. With some careful thinking and a little time spent on planning, we can make intelligent, healthy and environmentally friendly choices about the furniture we live with.

Futons

Futons are traditional Japanese mattresses made from layers of cotton wadding, either three or six held together in a cotton cover with ties to keep it stable. These have the advantage of being very hard and therefore good for the posture, as well as being adaptable pieces of furniture for the home. They can be made into sofas as well as beds by folding them together.

Synthetic Veneers

These are more commonly known as formica. The technical name for them is laminates or melamine formaldehyde. These are made from synthetic resins bonded together under high pressure onto Kraft paper. The resins are by-products of the meat industry. Laminates are very stable and do not break down very easily so do not emit gases into the home.

Taps

Make sure that taps can be used by all the family. The best type are those used in hospitals with levers rather than screw fittings. Check washers to make sure that your taps aren't dripping, and therefore wasting water and energy.

Toilets

A toilet uses 9 litres of water for each flush. There are toilets on the market which only use 6 litres. If you are not on mains drainage there is a toilet that only uses 3.75 litres of water. For the more dedicated there are composting toilets and electrically assisted aerobic toilets.

Wood Veneers

Veneer with chipboard and MDF is the most common form of veneer today. Most of the veneer comes from Belgium. Veneer is a relatively economical way of using tropical forest wood as 50 veneered table tops can be made from the wood that it would take to make one 25mm thick solid wood table.

However, we must ask ourselves, as concerned, informed consumers, whether we should be using this wood at all. Is it not better to use our own sustainable grown timbers for solid furniture, which will greatly outlast the veneered variety in any case?

11 Medicine

Introduction

We must bring the caring back to curing. Looking after ourselves has a direct and profound influence on the way we think about others and the planet we live on.

Penny Davidson

Being healthy in this fast-moving, stressful world can seem like an impossible dream. We are bombarded with 'short-cuts' to help us: vitamin pills; supplements; new healthier diets all claiming fantastic and almost miraculous properties. One of the new health 'cons' has been the upsurge in sales of various tonics and pills that are said to be the new elixirs of life, solving every single health problem we have ever had. New age alternative medicine has been revered by some and heavily criticized by others. Western doctors claim it has no scientific validity or justification, and newspapers wait to pounce on scandals of unhappy clients spending thousands on useless 'supercures'.

The truth is, however, that both conventional and new age medicine have their problems. Traditional health service doctors have only been around for the last hundred years, and people in many countries managed to get along without the single issue medicine they practise. Over £2 billion is spent each year in the United Kingdom on pills, over 90 per cent of that is on prescriptions, and yet most of them do no good whatsoever. In fact, the World Health Organisation tells us that only 200 of the 18,000 drugs available to doctors are necessary or useful, and yet they still continue to sell them, often testing them on animals first to see the effects.

Our Health – under whose Control?

It has become normal for us to believe that someone will look after our health, in the same way we have expected someone else to look after the health of the environment. That is why we trust doctors, as well as alternative therapists and healers, to solve our problems. There is no doubt that modern medicine has been beneficial to the vast majority of us. Antibiotics have meant that previously fatal illnesses can now be cured within a week or so, and using therapists to talk through emotional problems with individuals or families can save relationships as well as future health and stress-related problems.

Modern surgery means that we can fix up a broken leg, replace a heart or liver, remove a baby by Caesarian section, and replace the cornea of the eye. We can keep people living longer – but only to a certain point; in our rundown health service we now find that we can no longer afford the cost of keeping people alive for *too* long.

During the last hundred years we have to a large extent eradicated illnesses like smallpox and polio in industrialized countries, but have been unable to prevent the increase of others like cancer and heart disease.

Human Health – Planetary Health

The health of our bodies is directly connected to the health of our environment. The air we breathe, the food we eat, the type of housing we have all creates an environment that makes us feel good or bad and gives us a longer or shorter lifespan. We are now likely to live longer than previous generations as the result of the social revolution which has taken place in the last one hundred years. Most people are enjoying a better lifestyle and a cleaner environment. We have seen the introduction of better sanitation and water cleaning facilities. In fact these simple measures have done more for the health of the population in the United Kingdom than any of the wonder drugs combined. Ironically, our increasing development and use of chemicals, industry and energy have started to cancel out some of the benefits we gained.

Industrial activity generates millions of tonnes of waste each year. Some 300 million tonnes per annum of highly toxic waste is produced by industrial countries alone. This waste poses a health threat to human and animal life; it can affect small numbers of people, such as workers, or larger groups, including whole towns and cities. There is no easy way to get rid of hazardous wastes produced by industry. In the United Kingdom over 80 per cent of the industrial hazardous waste is simply landfilled – put in a hole in the ground. Nearly 10 per cent is dumped at sea and the rest is incinerated. The incineration of hazardous waste has come under intense criticism because of bad management practices and the subsequent air pollution that it causes, including the production of deadly dioxins. As we begin to criticize disposal of toxic or hazardous material near our homes or water supplies because of proven health problems, a new and more dangerous approach has been adopted by the major companies – shipping the wastes to Third World countries.

During 1986 over three million tonnes of hazardous waste was shipped *legally* to Third World countries. The amount of waste shipped by unscrupulous or illegal traders is unknown. Some countries have found that the toxic materials have simply been washed up on their beaches and others have found themselves deliberately misled about the exact nature and toxicity of the product. There is obviously anger from Third World countries about the dumping of wastes – why should they suffer because of our lifestyle?

Industrial accidents using hazardous materials are a daily occurrence, often releasing gases and fumes which can stay around for years, if not centuries. Accidents at nuclear power stations like Chernobyl represent only a small fraction of the numbers of serious and polluting accidents that occur. Literally millions of tonnes of dangerous materials are disposed of by industry every day, into our water systems, into the air, and the soil.

Severe health problems relating to air pollution obviously cut across international borders. In the United States over 150 million people breathe air that is considered unhealthy by the Environmental Protection Agency. In Greece's ancient capital, Athens, there is a significant increase in the number of deaths in days when the air is heavily polluted. In Bombay breathing the air is considered to be equivalent to smoking a packet of cigarettes each day, and air pollution is reckoned to kill one in seventeen people in Hungary.

All this pollution is of course directly related to our

lifestyles; industry doesn't create polluting wastes and toxic fumes for nothing. They are busy manufacturing all those chemicals for us to use in our homes and workplaces. It is our use of cleaners, machines, cars and synthetic materials which is responsible. The price we will pay for this over-consumption may well be our health. The car alone is responsible for over one half of all the world's air pollution, adding to global warming, contributing to acid rain, and producing noxious fumes so potent that children can suffer permanent brain damage, and adults can die with asthma-related illnesses, from doing nothing more dangerous than walking in the street!

One solution to the problems of pollution from the car is of course to give it up. But who in this world is going to take seriously the proposition that cars should be restricted? Environmentalists call for an alternative transport policy as one of the most important and significant ways that we can clean up our immediate surroundings, encompassing better public transportation, better design and accessibility of our major goods and services. Over 30,000 people die in the United States from secondary illnesses related to car pollution, and over one million are involved in serious car accidents. The health costs of this are enormous, far greater than the cost of providing alternative transport.

The water we drink is contaminated, groundwater and rivers are threatened by the use of pesticides, fertilizers like nitrates and industrial polluters, sewage and toxic waste. In poor countries the greatest difficulty is the enormous cost or bringing piped water to more people and the dwindling reserves of clean, safe ground water. Even water treatment plants cannot guarantee safe, perfectly clean water; all the chemicals cannot be fully removed and some of the cleaning and treatment chemicals themselves may be dangerous. More than half of the total population of industrialized countries uses water that has been treated. In London, water may have been recycled up to eight times.

Deforestation and desertification are pressing global concerns, but they represent the problems of industrializing and manipulating land and soil for the use of profits. The development of overgrazing, cash crops, use of chemicals and mechanization all lead to changes in our agricultural systems which are then reflected in the food on offer to the majority of people to eat. The subsequent health problems of poor diet, lack of fresh foods, loss of crops and poisoning of foods are not only problems for the industrial countries alone. We should be aware that what we eat affects every country we trade with.

Fishing is a good example of how health and well-being go hand in hand with a healthy environment. The overfishing and pollution of our major ocean fishing grounds has meant that the fish populations are dwindling. The fish themselves are subjected to pollution from our waste, and what was once seen as a source of healthy food is now no longer available in the quantities it once was. According to the United Nations Food and Agriculture Organisation, eleven major fishing grounds are at the point of collapsing. To overcome this shortage, fishing farms have been built which breed thousands of fish at a time and chemical additives have to be added to make the fish palatable to us. Farmed salmon, for instance, is coloured with canthaxanthin to make it turn pink so that it looks like fresh, wild salmon which feeds on a diet of shrimp, prawns and similar shellfish. Shellfish are often contaminated if they come from near a chemical plant or major pulp mill which uses chlorine. Some major fishing industries in western Canada have been closed due to dioxin contamination from paper pulping mills.

Attitudes to Health

Right at the very start of life, from conception to birth, our chances of survival are shortened by the food, lifestyle and housing of our parents. Environmental pollution, smoking, bad diet, stress and general ill health all affect the health of sperm or eggs. Our chances of survival and of good health depend on where we are born as well as how we are born. The poorer we are, the more likely we are to die earlier and suffer from various illnesses in the process. Environmental pollution, however, knows no boundaries; everyone is affected by air and water pollution (unless they are lucky or rich enough to be able to move).

The global health of our planet is therefore directly connected to the health of each individual. Our cancers, tumours, poisonings and so forth lead us back to an unhealthy *attitude* to our life, as well as an unhealthy life, and to an unhealthy planet.

Medicine

In industrialized countries the answer has been to provide services which attempt to cure the problems, without looking at the diseases themselves. Literally millions of dollars are being spent on cancer cures and similar wonder drugs, and very little on the solutions for the initial causes. Environmental and health conscious alternative groups expose problem after problem, with little obvious effect. While they might stop one chemical from being produced in one company, another company has already invented an alternative and it may well be several years before its health effects become known. The attitude has been: industry and superficial life first – and real quality of life second.

Positive Action for Holistic Health

1. Eat well; organic foods reduce your chemical intake and wholefoods help balance your diet. Eat less meat and reduce your sugar intake.

2. Take control of your own health first, let doctors and alternative therapists work with you to help maintain and cure your body.

3. Take regular exercise. Help your body take care of itself by exercising and movement, you will find you have more energy for other things, and better health.

4. Resist drugs; including alcohol, nicotine and tranquillizers. Resist the temptation to take every prescription from your doctor or to look for something to blank your mind from any worries of problems you might have.

5. Sleep and rest. Not enough of us care about getting enough sleep or relaxation, our modern lifestyles have encouraged us to be workaholics and fast-moving, to our detriment.

Directory

Acupuncture
Acupuncture works on the simple but effective principle of a balance between a positive and a negative force in the body.

Imbalances cause emotional and physical illnesses and the energy flows are released by gently touching small needles onto special points in the body. Acupuncture is a good cure for pain and can often help people with addictions.

Adrenalin Addiction

Working too hard and then not taking enough rest is a way to become addicted to adrenalin. We then place ourselves in situations in which we are stressed to produce more adrenalin, hence we only work when the deadline is close. Constantly worrying is another way to make ourselves stressed. This leads to burnout. Being unable to function when you get home because you are too tired is a sure sign. Find a pace that you can keep and remember that time to rest and relax is as important as working. Both are part of a cycle of renewal and action which keeps the body healthy.

Airborne Pollution

Sleep in a well ventilated room which has a good flow of air and is not too hot. This saves energy as well as being healthy. Still air, like still water, is a breeding place for bacteria. Do not sleep in a room in which someone has been smoking. In the home, the worst sources of air pollution come from smoking, wood treatments using synthetic air fresheners, mothballs, aerosol sprays and storing paints and solvents. Regularly open windows so that fresh air is circulated through the whole house.

Allergies

An allergy is a reaction of the immune system to an environmental pollutant. Allergies can be fairly innocuous, like sniffles or sneezes from grasses, or they may result in more serious illness, even death. Many people with allergies feel generally unwell and research has proved the link with long-term illnesses, like liver complaints and similar damage to major organs, with patients who have continuously found themselves susceptible to various allergic reactions. As we use more chemicals containing bioaccumulative poisons so the chance of allergies may increase. Try alternative medicines for relief from allergies, and cut down on synthetic chemicals generally.

Bathroom Cabinet

Take old drugs back to the chemist. Throw away over the counter drugs and replace them with a book on herbs and natural medicines. Use aloe cream for burns; calendula cream for cuts; arnica cream for bruises and knocks; olbas oil is very effective for colds; try gargling sage for a sore throat. The bathroom cabinet should have no more than the simplest of alternatives. Do not try to treat yourself for serious diseases unless you are properly trained.

Breastfeeding

It has been universally accepted that breastmilk is the very best food for babies, and has been so for millennia. Breastfeeding provides protection against allergies and against infections and offers babies the necessary complex immunizations they need to face the world. Bottlefeeding, on the other hand, has many documented risks, including contamination during manufacture or through carcinogenic nitrosoamines and the obvious contamination of the water supply. The manufacture and use of synthetic baby milk means that food is a major cause of the common forms of childhood diarrhoea; 200,000 deaths are recorded in Pakistan alone. Environmental problems include the waste packaging of millions of tonnes of metals and tins used for manufacture. Wherever possible, women should be encouraged to breastfeed their children.

Breathe Properly

Shallow breathing is a consequence of our sedentary lifestyle. Try to learn to breathe down into the stomach region of the body. As you breathe, you should be able to feel the stomach moving in and out with the breath. Deep breathing is essential for us to relax. It can relieve stress and is very beneficial to the body, as air is the start of the process which provides us with energy.

Cancer

One in five people in the industrial world is likely to be diagnosed with cancer at some point in their lives. The disease itself is largely caused by environmental pollution, lifestyle and diet, and is therefore potentially preventable. The human suffering and the cost to our nation are incalculable.

The most important thing you can do is to adhere to the positive health suggestions at the beginning of this chapter, and, of course, stop smoking.

Disinfectants
Many disinfectants contain trichlorophenol (TCP). Phenols are known to have caused death. Most wounds only need to be washed in hot water with soap and a little salt to clean the infected area. Serious wounds should be treated by a doctor.

Exercise
Regular exercise is very important for a healthy body and mind. If you have a job where most of the day you sit at a desk, then it is very important that you do some strenuous exercise at least once a week – more often if possible. Find the local swimming pool or gym and go there for a lunch time work-out. If possible, rather than taking the bus, walk. Walk up the stairs rather than taking the lift as an extra boost to the exercise programme. You will be surprised how quickly this will become an enjoyable activity rather than a chore, and you will feel better for it physically as well as save energy.

First Aid
Take a first aid course. This might be more helpful than buying lots of things in a first aid box that you do not know how to use properly. An alternative first aid kit could contain Rescue Remedy for shocks or tears, plasters and bandages, calendula creams for cuts and arnica for bruises, a basic first aid manual – the simpler the better – and any specialist creams or tinctures that you may need yourself. Keep the contents down to a practical minimum and store them safely out of reach of children. Of course, if you do not have chemicals in your first aid kit you won't need to worry quite so much.

Freedom from Clutter
Getting rid of clutter and rubbish from our lives can relieve us of boredom and tiredness. Tidy rooms or cupboards are so much easier to work with than messy ones. Most of us have far too much junk we could easily throw away (see Junk section, page 119). Throwing these things away, prefer-

ably recycling them, and being realistic about what we can and can't do is a very good way of relieving stress.

Herbalism

Growing your own herbs for healing minor illnesses is a wonderful way to help yourself. Herbs grow well in pots as well as in the garden. Start with lemon balm, which grows easily and makes a wonderful relaxing tea. Other easy-to-grow herbs are mint, which makes refreshing cold tea in the summer, sage for colds, and rosemary to add to a hot bath for cleansing and reviving. Cooking with herbs is a wonderful way to make food taste more interesting.

Curing yourself is very self-empowering, as you are using your own energy mixed with that of the earth. Many modern, post-war drugs are derived from herbs or from chemicals manipulated artificially to reproduce the same properties in natural herbs and plants. Herbs can help you to get well and get to know your body better. Try simple herbal remedies like mint to wake you up – brewed as a tea it makes a reviving drink. Don't try to treat major illnesses with herbs unless you have been trained, but most herbs will be able to be used with a good herbal guide without side-effects. Some, of course are deadly and need to left alone, so do read the warnings. David Hoffman's *Holistic Herbal* is one of the best guides along with the Potters guide, *Herbal Remedies, the Treatment of Common Ailments*.

Homoeopathy

Homoeopathy uses small amounts of herbs and minerals to stimulate our bodies' defences and healing mechanisms. The study of homoeopathy is complicated and individuals should not try and treat anything more than minor illnesses, cuts and grazes. As a form of healing it has little effect on the environment and is based on holistic medical practices.

Homoeopathic First Aid Kit

The six internal medicines listed below, together with the ointment, creams and tinctures, if promptly used, will often give quick relief, avert suffering and promote speedy recovery. Homoeopathic medicine is safe and free from toxic effects. Swallowing a whole bottle of tablets will do no harm – even to children.

The medicines
Aconite 6 Feverish colds. Chills. Sudden muscular pain from chills. Fright. Fear of coming events.
Arnica 6 Bruises. Any injury or accident. Physical or mental tiredness. Boils.
Arsen.alb 6 Mild food poisoning. Non-persistent diarrhoea. Vomiting. Running Colds.
Gelsemium 6 Influenza and influenza-like colds. Pre-examination nerves.
Nux Vomica 6 Indigestion. Effects of overeating or excessive drinking. Non-persistent constipation.
Rhus Tox 6 Lumbago. Sciatica. Relief of rheumatic pain. Aches and pains after excessive exertion. Sprains.

Dosage
Urgent 2 tablets every hour for six doses, then three times daily until better;
Less urgent Three times daily between meals until better;
Adults 2 tablets per dose.
Children 1 tablet per dose.
Infants 1 crushed tablet per dose.

Ointments and Creams (for external use only)
Arnica Cream for bruises;
Aloe Ointment for burns;
Hypercal Cream for cuts and sores.

Tinctures (for external use only)
Hypercal Tincture for cuts and abrasions;
Pyrethrum Liquid for bites and stings.

Hyperactivity
Many of our synthetic food additives can cause hyperactivity, especially in children, and should be avoided whenever possible. E numbers, the coding for various food additives and colourings given by the European Commission, list

those that cause hyperactivity including E102 (or tartrazine – a yellow/orange colouring) E107, E110, and E120. Hyperactive children need more time and attention at school than others; they can disrupt classes, suffer from overstimulation, and may eventually suffer from a wide range of illnesses including asthma, eczema, epilepsy and digestive disorders.

Immunization

A healthy body has its own immunization to fight disease and bacteria. A stressed, overloaded and sick body will find resistance to disease is low. We generally think of immunization as something we get in jabs when we are very small; but you can help your general immunization yourself by always eating a balanced diet, especially raw and fresh vegetables and fruits, and foods containing essential minerals. A less stressful lifestyle will obviously help you strengthen your immune system against disease.

Infertility

According to the World Health Organisation, infertility has now affected one in six couples in developed or industrial countries, and one in ten people worldwide. Environmental causes of infertility are wide-ranging. The build-up of cancer-causing chemicals and unnatural by-products in the environment particularly affects male sperm. Male sperm takes up to three months to mature and can be damaged by bad diet, alcohol, and pollution. Parents who wish to increase their chances of conception are advised to look after their bodies, particularly in the months before they wish to conceive, reducing any toxins they may be exposed to through food, drink, air and water. Penny Stanway's *Green Babies*, Random Century, 1990 has some good ideas on fertility and the environment.

Ionizers

The modern world has virtually eliminated all the negative ions from the atmosphere; electrical equipment and fluorescent lighting produce positive ions. Running water, however produces negative ions. Ionizers improve the quality of the indoor air that we breathe. By filling the atmosphere with negative ions, the dust particles in the air, which are positively charged, cling to surfaces. Breathing clean and pure air is a vital part of staying healthy.

Natural Medicine
Many people turn to natural medicine when all else has failed. Why wait till then? Like all cures, it relies on a belief that you want to get better. Illnesses can be your body's way of saying that something needs to change in your life. It may that you need to try a different sort of healing, or that you need to rest. The idea of a different approach to the way you get better could be the start of this process.

Noise
Noise is a major problem for many people; some have disturbed sleep, some people are made deaf by noise from industrial machinery at work. Noise is a pollutant. Traffic noise, for instance, can destroy the tranquility of a whole community. Try to lessen your own production of noise and you will find that your stress level drops accordingly.

Osteopathy
Osteopathy is medicine of the bones, spine and joints. Osteopaths believe that the body will be able to function better if the joints and spine are aligned and the structure of the body is balanced, and use manipulation of the joints to achieve this alignment. It uses no chemicals, testing or emotional management.

Pesticides
Good food is essential for good health. Pesticide contamination of food could be affecting as many as 250,000 people worldwide. Doctors know very little about pesticide poisoning and its treatment. The possible effects on the next generation are the most worrying, as pesticides seem to affect the genes. Organic food, grown without the use of pesticides is by far the healthiest option.

Population
The number of people on our planet is expected to grow to over 6.75 billion by the year 2000 and the population issue has become an increasingly important political and environmental subject. The problems, and the solutions, are not sim-

ple. Often families do not have access to family planning, women may be powerless, especially in religious or male-dominated societies. It is clear that support and education for women is a crucial factor in reducing the number of children born.

Posture
When sitting at desks or in chairs, make sure that you are sitting straight and that your spine is straight. Insist on having a desk that is the right height. At home have chairs that support the small of the back and the top of the neck. Sitting puts more strain on your back than standing. Sitting bent over more than doubles this pressure. To relax the best position is to lie down flat. In the home the best solution is to have large cushions and to lie around (roman style!).

Self Care
You are the best judge of your own health and well-being. Self-observation when you are feeling well and healthy is the best guide for the continuance of these feelings. Sometimes making the space to paint, or to do a hobby, will do more for your well-being than cleaner air or fresh food.

Sense of Smell
Our sense of smell is located in the primal part of the brain. It is very closely associated to the parts of the brain that control emotions and mood. Our noses are overloaded with smells from our surroundings, mostly chemical in manufacture. Do you remember the days magazines smelt of the print rather than cheap perfume? Use unperfumed products when you can. Better still, make your own. If you want a perfume in anything that you make, add some essential oil.

Silence
Noise pollution is an ever increasing problem. The body reacts to loud noise as a warning signal. As the world we live in becomes louder and louder, we have to cut out more and more sounds. The more we cut out the less that we can fully participate in life and living. Noise pollution is responsible for a wide range of health problems, including high blood pressure and headaches. In Europe people living near an airport are two or three times more likely to visit the doctor. Try

to have quiet times in your life when there is no interference from outside noise. Try to become aware of how sound affects you and how to find a way to remove undesirable noise. This can be done by sound-proofing rooms, or building hedges to block noise from the road. Sound is damped by the use of heavy furniture and thick wool carpets.

Sleep

We all need to sleep and rest, but many of us just don't get enough. Noise, stress, air pollution and various factors can keep us awake at night, or make those important hours seem of no use to our bodies. Sleeping time is very important; it is a time when our bodies can regenerate after a hard day's work or play, and our body tissue mends itself. To help you get a good night's sleep try meditation, drinking relaxing herb teas like camomile, and forgetting your worries and stresses, at least for the night. Create an atmosphere in your bedroom that is conducive to sleep and you will find it easier.

Smoking

Every day in the United Kingdom nearly 15 million smokers smoke their way through a quarter of a billion cigarettes. Smoking causes lung cancer and many other cancers, and heart disease, the United Kingdom's biggest killer. More than 100,000 deaths are caused directly from the effects of smoking each year. Illness from cigarette smoking is not just confined to the smokers themselves; passive smoking – when someone who doesn't want to smoke has to because of others – is also a killer, and the effects of passive smoking are now well known. Tobacco grown as a cash crop needs pesticides and fertilizers, and by continuing to smoke you will be encouraging further environmental pollution. Dioxins, the most deadly chemicals known, are also present in tobacco because of pesticide residues, and in the paper, which is bleached with chlorine. Many books and guides are now available to help you kick the habit before it kicks you.

Support

In holistic health terms, support is one of the crucial factors that can make the difference between good health and bad. Support while you are ill is not enough – many people don't get enough support while they are well and busy. Take time to assess your personal support mechanisms: do you have friends, practitioners and similar support when you need it?

Vitamin Pills
Vitamin pills are a very lucrative business, worth in the region of £135 million per year in the United Kingdom. Eat a healthy diet, take exercise and rest adequately, and you should not need these supplements. Vitamin supplements, though, might be necessary for smokers or those with an unhealthy lifestyle. Try to improve your life rather than the vitamin manufacturers.

Well Women Clinics
Women, always concerned about their family's health but often getting nothing more than a prescription for their problems are welcome at well women clinics. Women are invited to ask any questions they want about their own health, check facts and compare notes on diets, and also get cervical smear tests and breast examinations.

Yoga
Yoga is one of the oldest traditions of relaxation and exercise, and is highly valued as a preventive medicine. It reduces stress and anxiety, and creates a balance between mind and body. It costs little to learn and will be a skill you value for life.

12 Gardening

Introduction

A beautiful garden is a happy encounter.

Japanese Proverb

The garden is the easiest and best place to start green living. We have direct control over what we put into our gardens and how we would like them to develop. They can provide us with a source of organic vegetables and fruits, an aid to healthy eating which helps towards achieving a healthy body.

The garden is an area in which we can relax and get in contact with the the earth again. Many people think of the earth or the soil as being dirty. It does make marks on our clothes, but soil is the growth medium of the earth; without it we would not be alive. It provides some of the nutrients on which plants and other animals survive. In the soil are thousands of microorganisms. They help the breakdown of organic matter, so that the minerals and and other chemicals are available for the plants to feed on. Also contained within the soil are hundreds of insects; these help in the breakdown of organic substances as well. The soil is very much alive, and it represents a microcosm of the whole planet. We can make our gardens into miniatures of the world as we would like it.

Environmental Problems

A common environmental problem which develops in the garden comes from taking plants from the wild and replanting them away from their natural habitats. They may then be of no use to the wildlife in the garden, as very often the foliage and branches do not provide the right sort of habitat. One example of this misplanting are fast-growing hybrid conifers; their seeds do not provide food for the birds, and their thick undergrowth creates shade too dense for many plants to live in. This type of anomalous planting frequently occurs with foreign species imported into the United Kingdom; for instance, many bulbs that we plant for spring flowers, such as snowdrops, come from the wild in Mediterranean countries. Many of our native species are transplanted out of their proper habitat as well.

The use of pesticides in agriculture is an area of great concern because it poisons the earth and the water supplies. Of particular concern is the use of nitrates, which have now leached into the water table. As a result gardens are becoming the only safe places for wildlife to live, and wild animals are returning to live in cities, where foxes are often sighted, as well as kestrels and other birds of prey. Their numbers are rapidly diminishing in the wild because these animals are relatively high on the food chain. Insects living on vegetable matter inevitably absorb vast amounts of pesticides; they are then eaten by small mammals such as rodents. As we progress up the food chain we thus find higher and higher concentrations of pesticides, so that the large animals ultimately consume very large quantities of poison and die. This can happen in our gardens if we insist on using pesticides and poisons to kill insects: we could easily destroy the family cat or dog while trying to get rid of a few aphids.

It is only in the last sixty years that we have started to use so many pesticides. Originally spraying was intended to remove the moulds and lichens that affect many plants; this then left the insects that fed on the lichen nothing to live on but the leaves of the trees, rather than the moulds and lichen which grew on them. Other insecticides were subsequently needed to remove these insects. By making a break in the food chain a harmful cycle was started. Food chains are very complex and very easily tampered with; as soon as a food chain is broken an animal which once functioned naturally in its environment no longer fits in quite the way it should.

Another problem resulting from pesticide use is that diseases and pests are now building up resistance to the chemicals. The red spider mite, for instance, has developed an immunity to all the organochlorines and organophosphates and all attempts to control these insects now fail.

Our gardens have their own food chains, and if we work with nature we can grow the plants that we need (there will always be good years for one variety and bad for another, and this is perfectly normal). Humus, for example, gives plants resistance to disease. Rather than apply humus, however, potato growers have used chemicals to combat blight for a hundred years, and modern varieties may have become dependent on chemicals.

Many gardens discourage wildlife by paving over the earth and having no natural areas for the birds to live in. A wild, unkempt garden is heaven for animals and plants, and much better than a manufactured surburban garden where there is not a leaf out of place.

Benefits of the Garden

Producing some of our own food provides us with the satisfaction of being self-sufficient, as well as control over what we eat. Even if you have no garden, you can still grow some of your own food. Runner beans and tomatoes grow very well as pot plants in the house, as long as they are given plenty of water and kept in a sunny or at least southerly window (keep in mind that you will probably have to pollinate them yourself with a paint brush). It could be interesting to experiment with raising other vegetables in pots or window boxes as well; cucumbers, sweet peppers, and various types of salad greens, for example. You might want to look into installing a grow-light table or even a small greenhouse window. Many herbs do very well in pots, notably basil, rosemary, chives, parsley, thyme, and all the mints. Remember that potted plants often require slightly different care than those growing freely in the garden.

If you have no access to a garden where you live, you might be able to get an allotment. Allotments are pieces of land usually owned by the local council and leased to people. The library will have information about allotments in your area. Take advantage of this: a little research could bring you your own plot of earth where you can benefit not only from the fresh air and exercise but have the satisfaction of raising your own food! (Don't forget that if a plot of your own seems overly large or too time-consuming, you can always arrange to share with another would-be gardener.) Organic gardener John Jeavon estimates that it is possible to grow enough fresh produce for a year on an area of 10 square feet. New methods of intensive planting can yield heavy crops of high quality in small spaces; information on this can be found in organic gardening magazines and books. It is important to remember, however, that in small or city gardens we must be particularly careful not to exhaust the soil. Frequent applications of compost are necessary, and are an easy, natural and low-cost way of enriching the earth.

Weeds are a perpetual problem in the garden. One way we can avoid developing the habit of using herbicides is to review our attitudes towards these plants. While many weeds are, of course, invasive and threatening to the overall well-being of our gardens, others are harmless, often attractive to the eye, and can even be beneficial. Weeds, by shading the earth, can help to prevent soil dehydration during periods of drought, and their root systems can help to prevent erosion during heavy rains. Many people are starting to see the beauty of low-maintenance gardens filled with wildflowers and grasses which are otherwise known as 'weeds', such as wild cornflower, buttercups, daisies, dandelions, prince's feather.

There are many alternatives to chemical use for the control of weeds in the garden. One method which, in addition to being environment-friendly, is also creative and fun, is what we might call 'competitive' planting. In areas where invasive weeds have or might become a particular problem, for instance at the edges of flower beds or borders, create thick plantings of densely-growing groundcovers. Creeping plants such as certain varieties of ivy, periwinkle, phlox, thyme, loosestrife, heather and so on, and those which spread quickly or form tough clumps (forget-me-not, pinks, dwarf iris, poppies, pinks, violets) are excellent choices for keeping such persistent weeds as creeping charlie and bindweed out of your beds. Make your plantings at least one foot wide if you have the space, because many creeping weeds send out long runners which take hold very quickly. Those little cracks in the garden walk which can be so difficult to keep clear of weeds can be planted with ornamental mosses, or creeping thyme, which forms a low, springy cushion of highly fragrant foliage. Such intensive planting also helps conserve water, because less ground is exposed to evaporation.

If your lawn seems beyond hope, rather than spraying with herbicides or facing the prohibitive costs of having the area returfed, there are other choices. Often an abundance of weeds indicates an imbalance in the components of your soil. Creeping charlie, for example, thrives on earth with poor drainage and aeration. These conditions can be rectified without too much trouble or cost; consult a good organic gardening book or magazine for details (see the further reading section, page 193). Some weeds like soil which is particularly acidic or alkaline; these problems also can be easily corrected. Another possibility is to create an 'alternative' lawn. An original and low-maintenance 'lawn' can be made by having the turf removed and planting a mass of mixed groundcover. Such a lawn would need no mowing and could be in bloom from spring through autumn. A path could be made of flagstones, mosses, or creeping thyme. Or your lawn could be turned into an idyllic wildflower meadow. There are limitless possibilities.

Another method for natural weed control is mulching. The cost is minimal and, with a little initial planning and effort, you will be rewarded with a drastic reduction in the number of weeds in your garden. This in turn will spare you much labour and many hours in the future.

If you want to dig your soil, do it in the autumn, and remove as many weeds as possible, especially the roots (weeds, incidentally, can be placed on the compost so long as they have not gone to seed; if they have, you are simply asking for trouble next season!). In the spring, mix in a good feeding of compost, topping it off with a layer of compost at least three inches thick. When weeds do appear, they will be easy to remove from the light, loose soil, and your plants will thrive in this rich mixture. Once seedlings or bedding plants have been well established, apply more mulch and, preferably, a top layer of bark chips. The mulch provides excellent drainage, good aeration, rich nutrients, helps prevent soil compaction, keeps moisture from evaporating in hot weather, (thus conserving water) and discourages weeds. For the remainder of the season your garden will delight you with low-maintenance beauty and/or utility – chemical-free!

The popularity of organic gardening is growing rapidly as

public awareness and understanding of the pollution crisis increases. Consequently, many publications have become available which provide information for the concerned gardener. There are also large numbers of products on the market which provide an informed public with alternatives to the chemicals of a few years ago. One organic fertilizer company reports that sales of its products have risen by 52 per cent in the last year.

People are becoming more aware of the psychological benefits of having a garden in an ever-more hectic society. Studies have shown that those with illnesses recover more rapidly when they are recuperating within reach of gardens or parks. Vegetable gardens encourage self-sufficiency and ecologically positive and efficient use of land. All green spaces, but particularly those planted with vegetation and trees, help to fight global warming. Be part of the green wave: make your own garden.

Guidelines for the Greener Garden

1. Try to track down older varieties of vegetables, plants and flowers; they are often hardier than modern hybrids, and the vegetables frequently have better flavour.

2. Avoid all chemical pesticides and herbicides in the garden; use organic alternatives instead. Mulching, beneficial insects, intensive planting, and soap solutions are all possibilities to explore.

3. Grow groundcover plants and mulch the soil well, so that the garden can be watered less in the hot summer months.

4. Make your own compost and use along with manure, instead of commercial fertilizers.

5. Plant fruit trees for food, shade and to cool the air surrounding your home.

6. Consider letting your lawn become a groundcover garden or wildflower meadow.

7. If you have the resources, build a pond.

8. Encourage bees and butterflies by planting buddleia and other sweet smelling flowers. These creatures, along with other insects, songbirds, owls and bats, all help to maintain the natural balance of animal life within your garden.

9. Have nesting boxes of various sizes placed around the garden to encourage birds.

10. Make sure you have plenty of potted plants around: they renew and even filter the air in your home. Spider plants are especially good as they are very easy to care for.

Directory

Ants
Ants can be a nuisance in the home, though beneficial in the garden. A natural ant killer is a mixture of fresh chilli peppers, onions and garlic boiled in water and left to steep for several days. Add to a liquid soap solution and spray with a hand pump onto the ants.

Bats
Bats are a natural and necessary part of the garden. They prey on mosquitoes and other insects, and help to maintain the correct proportions of these creatures in the environment. Support your local bat.

Bees
Bees are essential for pollinating fruits and vegetables, and of course form a link in the food chain: other animals need them for food. They are ecologically very sensitive, and many wild bees have been wiped out by the extensive use of pesticides in recent years.

Bird boxes
These boxes have become essential because, through our continual building and urban development, we have

removed or altered the conditions necessary to support bird populations. Fewer trees and heavy pesticide use are only some of the most obvious factors threatening flying creatures; try to encourage owls and bats in particular, which have been very hard hit by the destruction of their habitat.

Bonfires
Bonfires of household products can cause toxic fumes when burnt, and all organic matter should be made into compost rather than being burnt or placed in the garbage.

Companion Planting
This is a natural way to encourage growth and reduce insect infestations. Some plants have chemical defences against pests, as they manufacture volatile oils which can repel insects. Some seem to grow better when they are close to other plants. Garlic and onions contain natural fungicides and germicides and should be planted throughout the garden (not only the vegetable garden, either: they produce very attractive blossoms and, oddly enough, a pleasant fragrance!). They seem to be especially effective in the rose garden. Herbs which are strongly aromatic act as insect repellents both inside and outside the home (be careful here, however: certain herbs may attract insects as well; for example, sweet basil is a gourmet dish for earwigs!). The table below is only a starting point; you may discover other combinations of plants in your gardening experiments.

Companionable Groups

Asparagus	calendulas and tomatoes
Beans	aubergines, leeks, marigolds, potatoes, rosemary, rhubarb, and strawberries
Beets	onions
Broccoli	aromatic herbs, onions
Brussels Sprouts	aromatic herbs, garlic, onions
Cabbage	celery, garlic, onions, peppermint, rosemary, sage, thyme
Carrots	chives, leeks, lettuce, and peas
Cauliflower	aromatic herbs, garlic and onions
Corn	marrows

Cucumber	corn, dill, radish and sunflowers
Garlic	roses
Leeks	carrots
Lettuce	broccoli, cabbage, radishes and shallots
Onions	cabbages
Parsley	asparagus, celery, leeks and peas
Parsnips	tomatoes
Peas	carrots and potatoes
Potatoes	garlic and horseradish
Radishes	lettuce and mint
Spinach	strawberries
Tomatoes	asparagus, basil, chives and parsley

Some combinations to avoid: brassicas (cabbage, cauliflower, broccoli, fennel, etc.) with marigolds, radishes or tomatoes.

Compost heaps

The most important part of the garden. Applying compost is the simplest and easiest way to enrich the soil. Recycling the organic waste from the kitchen into the garden will save time, money and energy, and help reduce landfill as well. (Almost 30 per cent of household waste is compostable.) All you need is a small bucket or container in the kitchen. Large glass jars – pickle jars, for example – are good, especially if you live in a flat and can't get out to your allotment very often. If you haven't any garden at all, you can still make compost and give it to your gardening friends!

It really couldn't be simpler: run a little bit of water into the bottom of your container, and then put in anything that will biodegrade: eggshells, leftover dairy products, and all types of vegetable matter. Autumn leaves and lawn clippings make very good compost, as do a few old newspapers. Do not place meat, bones or other such animal products on your compost. In the garden, compost can be placed in piles with boards or pieces of carpet over the top to conserve heat, or you can build a simple wooden composter with good ventilation through slats in the side and set a few inches off the ground. The composter must be able to 'breathe', but also keep in warmth.

Fertilizers

For the garden to be healthy it needs a rich soil. There are many ways in which you can feed your soil without using commercial fertilizers. Compost (see above) is always a good choice, and so is animal manure if left to rot before use. Hoof and horn, sometimes called bonemeal, and dried blood, are very good sources of nutrients; they are, however, animal by-products. Seaweed and potash can be bought from garden centres. Liquid plant manure can be made by putting the green leaves and stems of plants like comfrey and nettles into a water barrel and leaving them for at least a month. This produces a highly nutritious – and smelly! – plant food which can be used in watering, or sprayed onto the plants. Plants of the pea family are a very good source of nitrogen. The plants can be dug directly into the ground or put on the compost heap before they have gone to seed.

Garden Gnomes
These originate from the spirits of the earth. They were thought to be a representation of the way the plants suddenly seem to grow without any help.

Gardening for the Lazy
This takes some work at the beginning, but with careful planning and thought you can make a garden that, eventually, will virtually look after itself. The way to do this is to have a garden that has raised beds, lots of mulching and intensive planting. Raised beds are easier to cultivate and to harvest from and have fewer weeds; intensive planting further discourages weeds and requires less watering.

Goldfish
Goldfish in your pond will eat the larvae and eggs of creatures that breed in the water, and this will deprive the garden of natural predators and insects. However, goldfish are very useful in the rain water butts, as they eat the larvae of mosquitoes. In the winter remember to break the ice on the butt so that they can breathe.

Hedges
An important part of the garden; a natural hedge of beech or holly can provide a home for birds and other animals. Evergreen hedges, however, are too dense and dark to sup-

port much wildlife, and their foliage is too hard to provide compost for the garden.

Lanterns

These can be used to feature certain parts of the garden at night. Solar lights which store the energy of the sun during the day can light the garden at night. They are lightweight and very easily moved, so can be left in a sunny place during the day and moved to another spot for night illumination.

Lawns

Many environmentalists feel that lawns are a luxury and that the space should be given over to food production. This would save energy and resources because we would not have to import so much food (most organic food sold in supermarkets comes from Holland). Lawns do provide a pleasant green space around the home, but they should be kept to a minimum. You might consider retaining some traditional lawn but converting other parts to vegetable gardens or a meadow lawn which is only cut once a year. Lawns should be small enough to be cut by hand, thus reducing the need for a lawn mower, which uses energy and causes noise pollution. Weeds in the lawn can then either be dug out or left to grow. Lawns respond well to high nitrogen fertilizer. This can come from seaweed fertilizers or from compost containing peas and beans. Planning the site of your lawn can help to save on fertilizers; if the lawn is badly drained or heavily shaded it will never do well. Choose seed that suits your soil type. Besides wildflower lawns, a green space can be made with chamomile or periwinkle.

Money

This is often a worry, as most of us feel that we do not have enough. A well-planned vegetable garden will save money and is a good source of organic produce. One packet of seeds will produce more than enough plants to feed the whole family. Make friends with other neighbours who also garden and start up an exchange system where one person grows tomatoes, another courgettes and so on, and these are exchanged at harvest times. It is important to remember that organic gardening does not require expensive equipment or a large outlay of funds; on the contrary, careful thinking at the outset, energy, enthusiasm and a little ingenuity are far more important!

Naming
Naming your garden might help you to feel involved with and appreciative of it. In ancient times in China it was said that only when a garden has a name will its true splendour be seen. Works of art are given names, and our gardens can be our individual works of art.

Natural Flowers and Vegetables
These will grow as they do in nature; they do not have 'perfect' shape and colour. Postcard-pretty fruits and vegetables are grown with a lot of pesticides and chemical fertilizers.

Peat
This is recommended by many books and garden centres as a soil enhancer, yet it is an endangered natural resource: in the last fifteen years we have managed to dig up 96 per cent of United Kingdom peat bogs. These bogs were once the haven for many rare animals and birds. They cannot regenerate once they have been stripped because they are extremely delicate ecosystems. The compost that you make at home is a very good alternative to peat – spread the word!

Pesticides
Pesticides are part of an ongoing controversy surrounding their effects on our health and their effects on the environment. The evidence is clear: pesticides are toxic not only to plants and insects but also to humans. In one year the world produces 40 million tonnes of insecticides, herbicides and fungicides, 15 per cent of which is destined for domestic use. These chemicals stay inside the plants which we subsequently use as our food supply. While we continue to eat conventionally grown food there is no way of avoiding them.

Pesticides to avoid are:
- Arsenic compounds;
- Chlordane;
- Captan;

- Dichlorvos;
- DDT;
- Lindane;
- Mercury compounds;
- Parathion;
- Thiram.

Chemicals like these are being banned all the time, but they are quickly replaced by others just as poisonous. The chemical industry is very powerful and has a vested interest in our reliance on its products.

You can make your own natural pesticides for aphids: use a liquid of soap and garlic. Old washing up water is also effective. Nicotine, quassia, potassium soap and derris dust are also natural pesticides but these are sometimes highly poisonous substances of plant origin. They do break down in the soil, unlike chemical pesticides, but dangerous to have around the home, in case of accidents.

Planting

Plant lots of different varieties of plant; the greater the diversity of plant the more protection that they will each receive from pests and other predators.

Plant trees especially fruit trees, which have the advantage of being decorative feeding both wildlife and ourselves. Trees control the level of heat in the atmoshere. Think how cool and fresh it is sitting under and tree on a summers day. Well-placed trees near a house provide shade in the summer and reflect heat from the ground in the winter.

Ponds

Ponds in the garden attract frogs, toads and newts, all of which are important predators of slugs, snails and small insects. The easiest way to build a pond is to dig a hole and put in a pre-shaped pond-liner. These are more expensive and stronger than other liners. Fill it with water and plants. The insects will come of their own accord, or you can add a bucketful of water from a local pond. Frogs can be moved from friends' gardens, or some local conservation groups have lists of people with too many frogs and toads. As there are fewer and fewer ponds now in the wild, these animals need garden ponds for their survival.

Predators

These can be extremely beneficial in the garden and should be encouraged. Ladybirds control scales, mealybugs, thrips, mites and aphids. Orchard bugs control aphids, apple suckers, caterpillars, weevils, and eat the eggs of red spider. To attract benefical insects, make a mixture of honey, brewer's yeast and water in a old tin lid and leave in the branches of trees. Natural insect predators can be introduced to greenhouses to control pests. Encarsia controls whitefly for example.

Rainwater

Rainwater collection is a very good way of conserving water to be used in the dry summer months. There are also systems which use rainwater to irrigate the garden. In the summer it is worth diverting bath and washing up water into the garden. Baths can be drained using a hose and suction.

Self-Sufficiency in a Flat *is* Possible!

- Herbs can be grown in pots on window sills. Most require sun;
- Dwarf tomatoes can be grown on south facing windows; these take more care than the outdoor varieties;
- Lettuce can be grown in pots and troughs;
- Spring onions can be grown in window boxes;
- Runner beans in large pots on the floor;
- Mushrooms can be grown under the bed;
- Nasturtiums can be grown in hanging baskets (remember that both the flowers and the leaves can be used in salads).

Slug Pellets

Slug pellets are a danger in the garden. They leach their chemicals out into the ground if it rains and when the slugs die from the poison in the pellets they can be eaten by birds who ingest the poison. A natural way to kill slugs is to make up a solution of beer; water and sugar, and pour into an old tin lid placed in the ground, the lip being level with the top of the soil. The slugs will climb in and drown in the mixture.

Smells
Smells in the garden are a delight to our senses and to the bees. Strong smelling herbs and flowers are natural repellents to flies and other insects. Many also attract butterflies to the garden. Butterflies like other insects, have begun to die out because of the heavy use of pesticides. Pesticides kill all insects – not just the harmful ones.

Solar Greenhouses
A solar greenhouse built on the side of the house is not only a recreational area but provides solar heating indoors. It needs to be attached to the sun-facing side of the house, with the glass panels angled to get the maximum winter sun.

With careful planning, a greenhouse of 215 sq ft can grow 70 per cent of the salad vegetables and 30 per cent of the fruit needed for a family of four. Compost can be made from waste organic matter from the kitchen and converted into compost using a worm box. The amount of work needed is very little – perhaps an hour a day.

Tree Adoption
This is a very good way of growing trees if you do not have a garden. Many local parks and gardens have tree schemes, as do some of the large environmental organizations like BTCV, the British Trust for Conservation Volunteers.

Water Conservation
Only use water in your garden when absolutely necessary. The average US citizen overwaters their garden by up to 40 per cent. Protect the ground from water loss in the summer months by planting ground-covering plants or covering with straw. There are also products based on tree bark which can be used to do this. Another way is to put old newspapers several inches under the surface of the soil and cover with the top soil. This has the advantage of providing food for the insects that live in the ground, such as worms.

Window Boxes
These are decorative and can be used inside the house as insect repellents. Plant window boxes with herbs and other sweet smelling plants, basil is very good. When the sun is on them they will release their essential oils out into the house.

Flies do not like these smells and will avoid them. Lavender and geraniums are also good. Under windows plant nicotiana and other plants that release their scents at night.

Worms
Garden's greatest friend; they aerate the soil and improve soil quality by taking organic matter underground.

Further Reading

2 Minutes a Day for a Greener Planet, Marjorie Lamb, HarperCollins, 1990
1,001 Supersavers, Pamela Donald, Piatkus, 1990
1,001 Ways to Save the Planet, Bernadette Vallely, Penguin, 1990
4,000 Things You Really Ought to Know, Ginette Chevallier, Guild Publishing, 1990
All New Hints from Heloise, Perigree, 1989
The Bobbin Lace Manual, Geraldine Stott, Batsford, 1988
Cheaper and Better, Nancy Birnes, Optima, 1988
Classic Crafts, Martina Margetts (ed.), 1989
Cleaning and Stain Remover, Barbara Chandler, Ward Lock, 1988
Collage, Golden Hands, 1990
The Complete Book of Self-Sufficiency, John Seymour, Corgi, 1978
The Craft of Hand Spinning, Eileen Chadwick, Batsford, 1981
Creative Crafts, Guild Publishing
Electricity Council Literature
Energy Saving with Home Improvements, Which Books, Hodder & Stoughton, 1986
Fast Food Facts, Tim Lobstein, Camden Press, 1988
The Forgotten Arts, John Seymour, Dorling Kindersley, 1984
Friends of the Earth Handbook, Optima, 1990
The Good Wood Guide, Friends of the Earth, 1990
The Green Consumer's Guide, John Elkington and Julia Hailes, Gollancz, 1988
The Green Consumer's Supermarket Shopping Guide, John Elkington and Julia Hailes, Gollancz, 1989
Green Design, Avril Fox and Robin Murrell, Archetecture, Design & Technology Press/Longman, 1989

Green Living Magazine, Women's Environmental Network, 1989

Green Pages, John Button, Optima, 1988

Green Magazine Guide to the Home, Green Magazine, 1990

Healing Environments, Carol Veniola, Celestial Arts, 1988

Home Ecology, Karen Christensen, Arlington, 1989

How to be Green, John Button, Century Hutchinson, 1989

How to do Just About Anything, Reader's Digest, 1988

Introduction to Batik, Heather Griffin, 1989

An Introduction to Calligraphy, George Evans, Apple, 1987

Penguin Book of Kites, David Pelham, Penguin, 1976

Laura Ashley Complete Guide to Decorating, Guild Publishing, 1989

Living Style: Stencilling, Helen Barnett and Suzy Smith, Ward Lock, 1990

Make Your Own Jewellery, Everitt, Usborne, 1987

Mrs Beeton's Book of Household Management, Chancellor Press, 1989

The Natural House Book, David Pearson, Octopus, 1989

The New E for Additives, Maurice Hanssen, Thorsons, 1987

The Non-Toxic Home, Debra Lynn Dadd, St Martin's Press, 1986

Origami, Zulul Aytine Scheele, Hamlyn, 1990

Pay Less Keep Warm, Barty Phillips, Optima, 1987

The Politics of Food, Geoffrey Cannon, Century, 1988

Recycling: A Practical Guide for Local Groups, Waste Watch, 1990

The Self-Sufficient Kitchen, B & R, Macmillan, 1980

A Self Sufficient Larder, Mike Foxwell, Optima, 1988

The Smart Kitchen, David Goldbeck, Ceres Press, 1989

The Survivor House, David Stephens, 1988

Thorsons Organic Consumer Guide, David Mabey, Alan Gear, and Jackie Gear, Thorsons, 1990

You and the Environment, Which Books, Hodder & Staughton, 1990

Your Home, Your Health and Well Being, David Rousseau et al, Ten Speed Press, 1988

Address List

General

The Body Shop
 International Plc
Hawthorn Road
Wick
Littlehampton
West Sussex BN17 7LR

British Trust for
 Conservation Volunteers
36 St Mary' Street
Wallingford
Oxfordshire OX10 OEJ

Common Ground
45 Shelton Street
London WC2H 9HJ – (071)
 379 3109

Friends of the Earth
26–28 Underwood Street
London N1 7JQ – (071) 490
 1555

Greenpeace UK
Greenpeace House
Canonbury Villas
London N1 2PN – (071) 354
 5100

London Ecology Centre
45 Shelton Street
London WC2H 9HJ

OXFAM
272 Banbury Road
Oxford
OX2 7DZ – (071) 585 0220

The Women's
 Environmental Network
287 City Road
London EC1V 1LA – (071)
 490 2511

Building, Design and DIY

Auro Paints
16 Church Street
Saffron Waldon
Essex CB10 14W

Auro Pflanzenchemie
 GmbH
Postbox 1229
D–3300 Braunschweig
West Germany

British Gypsum
(lime plaster, gypsum)
Westfield
360 Singlewell Road
Gravesend
Kent DA11 7RY

The Builders Merchants
    Federation
15 Soho Square
London W1V 5FB – (071)
    439 1753

Bulmer Brick & Tile
Bulmer
Nr. Sudbury
Suffolk

Bush Masheder Associates
16 Church Street
Saffron Walden
Essex CB10 1JW

Cork Industry Federation
c/o Mrs J. Iliffe
62 Leavesden Road
Weybridge
Surrey KT13 9BX

Crucial Trading
(Quarry tiles)
PO Box 689
London W2 4BX

Ecology Building Society
18 Station Road
Cross Hills
Keighley
West Yorkshire BD20 7EH

Fired Earth
Twyford Mill
Oxford Road
Adderbury
Oxfordshire OX17 3HP

Gaia Associates Ltd.
115 Lower Baggot Street
Dublin 2
Eire

Gaia Environments Ltd.
Umbrella Studios
12 Trundle Street
London SE1 1QT

Glass & Glazing Federation
44–48 Borough High Street
London SE1 1XP – (071) 403
    7177

"Green Roofs"
Erisco Ltd.
Broughton Ho
Broughton Road
Ipswich
Suffolk IP1 3QS

Institute of Building Biology
16 Church Street
Saffron Walden
Essex CB10 1JW

Keith Critchlow
Royal College of Art
London SW7

Keystone Architects &
    Designers Cooperative
52 Gladsmuir Road
London N19 3JU

Address List

London Architectural
  Salvage & Supply
  Company
Mark St
off Paul Street
London EC2

Natural Paint Company
101 Lansdowne Road
London N17

Findhorn Foundation
The Park
Findhorn
Forres
Grampian IV36 OT2

The Survivor House
The Solar Housing Club
Victoria House
Bridge Street
Rhayader
Powys LD6 5AG

Swedish-Finnish Timber
  Council
21 Carolgate
Retford
Nottinghamshire DN22 6PZ

Wicanders GB Ltd.
Stoner House
Kilnmead
Crawley
West Sussex RH10 2BG

Power and Energy

Association for the
  Conservation of Energy
9 Sherlock Mews
London W1M 3RH

Atkins Building South (128)
Campden Hill Road
London W8 7AH

British Wind Energy
  Association
4 Hamilton Place
London W1V 0BQ

Campaign for Lead Free Air
(CLEAR)
3 Endsleigh Street
London WC1H 0DD

Centre for Alternative
  Technology
Llwyngwern Quarry
Machynlleth
Powys SY20 9AZ

Communities Against
  Toxics
129 Edison Way
Hemlington
Cleveland TS8 9ES

Confederation for
  Registration of Gas
  Installers
St Martins House
140 Tottenham Court Road
London W1P 9LN – (071)
  387 9185

Draught Proofing Advisory
  Association
PO BOX 12
Haslemere
Surrey

Ecological Design
  Association
20 High Street
Stroud
Gloucestershire GL5 1AS

Energy Efficiency Office
Department of Energy
Thames House
South Millbank
London SW1P 1LF

Energy Efficiency Service
Department of the
  Environment
Room 312 Thames HO
  South
Millbank
London SW1P 3QJ

Energy Research Group
Open University
Walton Hall
Milton Keynes MK7 6AA

Environmental
  Investigations Limited
Netley House
Gomshall
Surrey GU5 9QA

Filsol Ltd.
Ponthenri Industrial Estate
Ponthenri
Dyfed SA15 5RA

Heating & Energy Saving
  Centre
The Building Centre
26 Store Street
London WC1E 7BT – (071)
  637 1022

Heating & Ventilation
  Contractors Association
34 Palace Court
London W2 4JG – (071) 229
  2488

Institute of Plumbing
64 Station Lane
Hornchurch
Essex RMN 6NB – (04024)
  72791

International Solar Energy
  Society UK
Kings College
London

The National Radiological
  Protection Board
Chilton
Didcot
Oxfordshire OX11 0RQ

Practical Alternatives
Victoria House
Bridge Street
Rhayader
Mid Wales LD6 5AG

Marlin Lighting Ltd.
(Energy-saving lights)
Feltham
Middlesex TX13 6DR

Solar Energy Unit
University College
Newport Road
Cardiff CF2 1TA

Solar Trade Association Ltd.
Brackenhurst
Greenham Common South
Newbury
Berkshire RG15 8HH

Sungro-Lite Ltd.
(UV-emission-free lights)
118 Chatsworth Road
Willesden Green
London NW2 5QU

Truelite SML
(Full-spectrum lighting)
Unit 4
Wye Trading Estate
London Road
High Wycombe
Bucks HP11 1LH

Warmer Campaign
83 Mount Ephraim
Tunbridge Wells
Kent TN4 8BS

Wotan Lamps Ltd.
1 Gresham Way
Durnsford Road
London SW18 8HU

Flooring and Fabrics

Black Sheep
(Untreated and Vegetable-
  dyed woollens)
9 Penfold Street
Aylsham
Norfolk NR11 6ET

The British Carpet
  Manufacturers
  Association
Royalty House
72 Dean Street
London W1

Celeriac Whitley Willow
  Mills
Lepton
Huddersfield HD8 0NH

The Crafts Council
(Naturally dyed fabric)
12 Waterloo Place
London SW1Y 4LN

Futon Company
82/3 Tottenham Court Road
London W1P 9HD

Limericks
(Cotton blankets)
Limerick House
117 Victoria Avenue
Southend-on-Sea
Essex SS2 6EL

Mid-Essex Trading
  Company
Montrose Road
Dukes Park Industrial
  Estate
Chelmsford
Essex

Natural Fibres
(Mail order)
2 Springfield Lane
Smeeton Westerby
Leicester LE8 0QW

Peter Reed Textiles
Springfield Mill
Churchill Way
Lomeshaye
Nelson
Lancashire BB9 6BT

Cleaning

Ecover
Full Moon
Charlton Court Farm
Mouse Lane
Steyning
West Sussex BN4 3DF

Henry Flack
PO Box 78
Beckenham
Kent BR3 4BL

The Little Green Shop
8 St George's Place
Brighton BN1 4GB

Food and Drink

The Soil Association
86 Colston Street
Bristol BS1 5BB

The Vegan Society
7 Battle Road
St Leonards on Sea
East Sussex TN37 7AA

Equipment

Compost Toilets
(Mains shower systems)
Swedal Leisure (UK) Ltd.
PO Box 14
Egham
Surrey TW20 0QP

Mantaleda Bathrooms
(Showers, saunas,
    whirlpools & deep-soak-
    ing tubs)
6–10 Progress Row
Leeming Bar Industrial
    Estate
Northallerton
North Yorkshire DL7 9DH

London Ionizer Centre
(Ionizers)
65 Endell Street
London WC2H 9AJ

Junk and Waste Disposal

Aluminium Can Recycling
    Association
Suite 308
1–Mex House
52 Blucher Street
Birmingham B1 1QU

Federation of Childrens
    Resource Centres
c/o 25 Bullivant Street
St Ann's
Nottingham NG3 4AT

Pre-School Playgroups
Association
61–3 Kings Cross Road
London WC1X 9LL

Hobbies and Home Crafts

Brobie & Middleton
(Pigments)
68 Drury Lane
London WC2B 5SP

The National Association of
Paper Merchants
Hamilton Court
Gogmore Lane
Chertsey
Surrey KY16 9AP

Wiggins Teape Paperpoint
130 Long Acre
London WC2E 9AL

Furniture

Alphabeds
8 Foscote Mews
London W9 2HH – (071) 289 2467

Alternative Sitting
(Back-care furniture)
PO Box 42
Abingdon
Oxfordshire OX14 2EH

Atelier Furniture
(Furniture from recycled timber)
Glenholm
Geroge Street
Nailsworth
Gloustershire GL6 0AG

The Back Shop
24 New Cavendish Street
London W1

The Back Store
330 King Street
London W6 0RR

Balans Chairs
Lecton House
Lake Street
Leighton Buzzard
Bedfordshire LU7 8RX

Community Furniture
Network
Highbank
Halton Street
Hyde
Cheshire SK14 2NY

Dunlopillo
(Beds and bedding)
Pannal
Harrogate
North Yorkshire HG3 1LJ

Full Moon Futons
20 Bulmershe Road
Reading
Berkshire RG1 5RJ

German Bedding Centre
138 Marylebone Road
London NW1

Glass Manufacturers
   Association
19 Portland Place
London W1N 4BH

Hope Mill
112 Pollard Street
Manchester

Medicine

Neals Yard Remedies
5 Golden Cross Walk
Cornmarket Street
Oxford OX1 3EU

Gardening

Henry Doubleday Research
   Association
National Centre for Organic
   Gardening
Ryton-on-Dunsmore
Coventry CV8 3LG

Permaculture Association
8 Hunters Moon
Dartington
Totnes
Devon TQ9 6JT

Suffolk Herbs
Sawyers Farm
Little Cornard
Sudbury
Suffolk CO10 0NY

Mail Order

Cosmetics To Go
29 High Street
Poole
Dorset BH15 1AB

Traidcraft
Kingsway
Gateshead
Tyne & Wear NE11 0NE

The Whole Thing Catalogue
Millmead Business Centre
Millmead Road
Tottenham Hale
London N17 9QU

Index

acid rain 29, 31-2
acupuncture 166
additives *see* food
adhesives 49-50
adrenalin addiction 166
advertising 76, 106
aerosols 77, 114
air conditioning 34
air fresheners 74, 77
airing clothes 77-8
alcohol 92
allergies 166-7
allotments 178
aluminium 52
animal testing 73
ants 182
appliances 34, 103-18, 126-7
appliqué 137
aquariums 137
architecture 17-21, 23-4
asbestos 50
asphalt 52
auctions 121, 123

bamboo 70, 154
basketry 138
bathroom cabinet 167
baths 34-5, 78, 154-5
batik 138
bats 182
batteries 35, 107, 108
bedroom furniture 108
beds 108, 155
beer-making 138
bees 182
 keeping 135, 138
beeswax 155
bicarbonate of soda 78

bioclimate 20
bird boxes 183
bleaches 63-4, 74, 75
blockboard 52, 156
boilers 35
bonfires 183
books 124
boot sales 121
borax 78
bottle banks 124-5
bottling 92
bran 78
brass 79
bread-making 133-4, 138-9
breast-feeding 167
breathing 167
bricks 50, 52
brooms 79
buildings 17-28
bulk-buying 92, 93

calligraphy 139
cancer 167-8
candle-making 139
carbon dioxide emissions 29, 31, 32, 105
cardboard boxes 125
carpets 59-61, 64
 cleaning 79
cash crops 92-3
cavity walls 35-6, 48, 50
ceilings 36
cement 51-2
CFCs (chlorofluorocarbons) 47, 105, 108, 110, 148
charity shops 121, 125
chemicals
 in cleaning products 73-6

in house-building 46-7, 48
  safety guidelines 55-6
  see also pesticides
children 10
chimneys 36
chipboard 48, 53, 156, 157
chocolate 93
cleaning 73-88
  environmental effects of 73-6
  guidelines for buying 76-7
  products 73-7
  shopping list for 88
clocks 108-9
clothes
  airing 77-8
  brushing 79
  recycling 121, 125
clutter 168-9
coastal flooding 31
Cochrane, Josephine 105
coffee 93
coir 64
collages 125-6, 139
community databanks 136
companion planting 183-4
company policies 12, 15
composite boards 156
compost 109, 126, 184
concrete 50-1
consumerism 119-20
consumers, power of 12
containers 79, 110
convenience foods 93-4
cookers 109
cooking 37
  see also food
co-operatives 11, 93
copper 79
cork 53, 64
cotton 62-3, 64-5
crafts 133-47
creosote 51
curtains 37
cutlery 79

dampness 22
dandruff 79-80
decorating 45-58
deforestation 95, 149-51, 163-4
descalers 75
desertification 163-4
dieting 94
direct mail 128
dishwashers 105, 109
  powder 80
disinfectants 168
do-it-yourself 45-58
  materials 52-4
doormats 80
doors 37, 156
double glazing 37-8
drains 80
draught-proofing 37, 38
dress-making 140
drink 89-102
drugs 160, 165, 167
dry-cleaning 65, 80-1
dustbins 81, 109, 126
dyes 65-6, 140

eating out 94
education 11
eggs 94
electric blankets 109-10
electric cables 22
electric toothbrushes 110
electrical goods *see* appliances
embroidery 140
energy 29-44
  consumption 29, 32-4
  efficiency 22-3, 32-4
  environmental effects of 29-32
  sources of 29, 30-1
  see also solar; wind power
environmental awareness 12
environmental groups 11
equipment 103-18
  guidelines for buying 107-8
eucalyptus 69-70, 81
exercise 168

fabrics 59-72
  conditioning 81
  finishes 66-7
  flame-resistant 67
  natural 62-3
  non-organic 69
  synthetic 61-2, 69, 71
factory farming 90, 95, 98
fast food 94-5, 128
fat 95
feathers and down 67
fertilizers 185
fibreboard 53, 156
fibres, recycled 54
fillings 157
fire 51, 67
  extinguishers 111

Index

first aid 168, 169-70
fish 95, 164
flat-living 189
flooring 59-72
floors
    insulation 23, 38
    joists 51
    polish 81
    synthetic 71
    wooden 72
flower-arranging 140-1
fly killers 81-2
fly-tipping 14, 131
food 89-102
    additives and pollutants 91-2, 170-1
    bulk-buying 92, 93
    consciousness 89-90
    convenience 93, 94-5
    ethics 90-1
    labelling 91, 92, 97
    organic 91, 99
    poisoning 95-6
    preserving 92, 104
    raw 100
    safety 89
    steaming 100
food processors 106, 110
formaldehyde 48, 67, 149
formica 158
fossil fuels 29, 30, 31
frames 157-8
freezers 104, 108, 110
fruit stains 82
fungicides 49, 58
furniture 127, 148-59
    design 148-9
    guidelines when buying 154
    polish 82
    restoring 122
futons 158

gadgets 103, 105-6, 110-11
garden gnomes 185
gardens 176-91
    benefits of 178-81
    costs 186-7
    design 23
    furniture 127
    guidelines for 181-2
    for the lazy 185
    naming 187
    organic 179, 181, 186-7
    planting 187, 188
    in pots 178, 189
    smells 190
    tools 111
gas 30, 38
Gaudi 17
gelatine 96
glass
    engraving 141
    recycling 124-5
glasses 111
global warming (greenhouse effect) 29, 31, 74, 77, 150
glues 49-50, 149
glycerine 82
goldfish 185
grants 38
grass stains 82
greenhouse effect *see* global warming
greenhouses 190

hairbrushes 82
halon 111
hard water deposits 82
Hardyment, Christina 103
haybox 38-9
health 160-75
    action plan for 165
    and environment 161-5
heat exchanger 39
heating
    central 35, 36, 38, 39
    solar 24-5, 26-7
hedges 185-6
hemp 67
herbalism 169
herbs 96, 183
hobbies 133-47
Holden, Edith 147
homes
    building and design 17-22, 23-4
    environmental effects of 19-20, 23-4, 46-7
    planning 12-13
    self-built 19, 21
    'sick' 27
    traffic flow through 27
homoeopathy 169-70
honey 96
    see also bees
hot water bottles 109
human rights 11
humidifiers 39, 111
humus 178

hyperactivity 170-1

immunization 171
infertility 171-2
information, access to 15
ink stains 82
insulation 32
   cavity wall 35-6, 48, 50
   ceiling 36
   curtains 37
   floor 23, 38
   grants 38
   loft 40-1
   materials 48
   water tanks 43-4
   window 37-8
ionizers 39, 111, 171
ironing 83, 111
irradiation 96-7

jam-making 141
jars 112, 127
Jeavon, John 179
jewellery 141
jumble sales 121, 127-8
junk 119-32
junk food *see* fast food
junk mail 128
jute 68

kapok 68
kettles 83, 112
kite-making 141-2
knitting 142
knives 112

labelling
   for chemicals 51
   for energy consumption 39-40
   for food 91, 92, 97
lace-making 142
lanterns 186
latex 70
lawns 180, 186
lead 52
leather 68
leatherwork 142
leaves 128
lemon juice 83
letter-writing 12
lifestyles 9-10
lighting 25, 40
limestone 53
linen 68, 87

linoleum 61, 68
litter 128-9
lofts 40-1

Mackintosh 17
marble 53
marbling 143
markets 97
meat 95, 97-8
medicine 160-75
   natural 172
Melville, Arabella 94
metals 53, 157
   recycling 129
meters 41
microwave ovens 94, 112
mildew 54
milk 98
Moore Lappe, Frances 89-90
mops 83
moth balls 68-9
mulching 180
murals 143
muslin bags 112

newspapers 129-30
nitrates 98, 177
noise 23, 44, 113, 172, 173-4
nuclear power 30-1
nuresry equipment 112-13

oil 130
olive oil 98-9
onions 83
organic produce 91, 99
origami 143
osteopathy 172
oven cleaners 83
ozone depletion 47, 75, 105

packaging 99, 119, 120
paint 49, 52, 54-5
   removers 55
pans, burnt 83
paper
   making 143
   recycling 122, 129
papier mache 143-4
parenthood 10
parquet flooring 69
particleboard 53
party equipment 113
patchwork 144
peat 187

Index

pens 113
percolators 113
perfume 173
personal stereos 113
pesticides 60-1, 62-3, 69, 91, 101, 172, 177-8, 187-8, 189
pewter 84
phosphates 74
plaster 55
plasterboard 53
plastics 130
playgroups 130
Plplawski, Stephen 106
plumbing 24
plywood 54, 156, 157
poisons 51
polishes 81, 82
politics 14-15
pollution
    air 162-3, 166
    from energy sources 29-32
    indoor 19-20, 27, 48
    noise 23, 44, 113, 172, 173-4
    oil 29
    water 73, 163, 164
ponds 188
population 172-3
porches 41
posture 173
pottery 144
predators 189
pressure cookers 94, 114
printing 144
pulses 99-100

quarry tiles 84
quilting 144-5

radiators 41
radios 114
rainwater 189
    barrels 114
rattan 70, 154
raw food 100
rayon 69-70
razors 114
recycling 14, 99, 109, 119-23
reeds 70
refrigerators 104-5, 108, 114
relationships 10
remote controls 114-5
restaurants 94
roofs 24-5
    grass 24

    solar 24-5
    treatments 24, 28
rooms, placement of 25
rubber 54, 70
rugs 84
    making 145
rush matting 70
rust 84

safety
    and chemicals 55-6
    and food 89
salmon 95, 164
salt 84
saucepans 115
schools 130
scourers 84
scrap-stores 130-1
seagrass 70-1
seasons
    and design 25
    and food 100
self care 173
self sufficiency 189
septic tanks 56
shampoo 79-80
shareholding 12, 15
shellac 56
showers 34-5, 115
shutters 25, 37
silence 173-4
silk 71
silver 84-5
skips 131
slaughtering 90
sleep 174
slug pellets 189
smell, sense of 173, 190
smoking 174
soap 85
solar
    cell 42
    collector 41-2
    energy 26, 41-2, 107
    greenhouse 190
    heating 24-5, 26-7
    roofs 24-5
    water 26
solid fuel stoves 42-3
solvents 56-7
spectacles 131
spiders 85
spinning 145
sponging 145

sprouts 100
stain removal 71, 75, 82, 87
static 64
steamers 115
steaming food 100
Steiner, Rudolf 17, 18
stencilling 145-6
stone 54
sugar 100-1
sunbeds 115
support 174-5

tablewear, disposable 113
talcum powder 85
Tant 18
taps 158
tea 101, 106
    leaves 85
teapots 115-16
teflon-coated utensils 116
telephones 116
televisions 115, 116
textiles *see* fabrics
thermos flasks 85, 116
thermostats 43
Third World 9-11, 135
timber *see* wood
tipping *see* fly-tipping
toasters 116-17
toilets 86, 158
tool kit 57-8, 124
toothbrushes 110
toothpaste 86
toxic waste 131-2, 162-3
toy-making 146
training 11
transport 163
trees 47-8, 149-54, 188
    adoption 190
    see also wood
tropical rainforests 150-1
    see also deforestation
tumble driers 117

vacuum cleaners 117
vacuum freshener 86
varnish 52
veganism 101
vegetables 101, 187
vegetarianism 94, 101
veneers
    synthetic 158

    wood 158-9
ventilation 27, 43
video recorders 117
vinegar 86
vitamin pills 175

wallpaper 48-9, 58
Waring, Fred 106
washing lines 117
washing machines 117-18
washing powder 73-4, 86-7, 118
washing-up liquid 87
waste 14, 119-32, 162-3
waste disoposal units 118
waste treatment 120
water
    filters 118
    saving 190
    softener 87
    solar 26
    tanks 43-4
weaving 146
weeds 179-80
Well Woman Clinics 175
whitening linen 87
whitewash 58
wholefoods 101-2
wicker baskets 118
wildlife 176-8
wind power 27-8, 44
window boxes 189, 190-1
windows 37-8, 44
wine stains 87
wood 149-54
    endangered 154
    floors 72
    hardwoods 47-8, 108, 152-3, 157
    softwoods 153, 157
    sustainable sources 108, 150
    treatments and preservatives 28, 47, 57, 58
    veneers 158-9
    working with 146-7
wool 60-1, 67, 72
    sources of 72
workplace 11
worms 191
writing 147

yoga 175

zinc 87